MULTIVARIATE ANALYSIS

A Selected and Abstracted Bibliography, 1957-1972

STATISTICS

Textbooks and Monographs

VOLUME 1: The Generalized Jackknife Statistic
H. L. Gray and W. R. Schucany

VOLUME 2: Multivariate Analysis
Anant M. Kshirsagar

VOLUME 3: Statistics and Society
Walter T. Federer

VOLUME 4: Multivariate Analysis:
A Selected and Abstracted Bibliography, 1957 - 1972
Kocherlakota Subrahmaniam and Kathleen Subrahmaniam

OTHER VOLUMES IN PREPARATION

MULTIVARIATE ANALYSIS

A Selected and Abstracted Bibliography, 1957-1972

Kocherlakota Subrahmaniam and Kathleen Subrahmaniam

Department of Statistics
University of Manitoba
Winnipeg, Manitoba, Canada

MARCEL DEKKER, INC. New York 1973

MARCEL DEKKER, INC.

95 Madison Avenue, New York, New York 10016

LIBRARY OF CONGRESS CATALOG CARD NUMBER: 73-90690

ISBN: 0-8247-6190-1

PRINTED IN THE UNITED STATES OF AMERICA

TABLE OF CONTENTS

FOREWORD

We first became aware of this bibliography in August, 1973, after it was already completed. The structure seemed very good and it also appeared to be in a form which would make for easy use. The decision to publish was difficult because of the recently published "A Bibliography of Multivariate Statistical Analysis" by T. W. Anderson, Somesh Das Gupta and George P. H. Styan (Oliver & Boyd, Edinburgh, 1972). The Anderson-Gupta-Styan bibliography covered papers through 1966 while this bibliography includes abstracts of papers through 1972. The Anderson-Gupta-Styan text covers all of multivariate analysis while this one is limited to normal and related distributions. The Anderson-Gupta-Styan text classifies articles while this text also gives abstracts of the selected papers. This text starts with the assumption that the textbook, "An Introduction to Multivariate Analysis" by T. W. Anderson is available for the researcher and builds on the bibliography in Anderson's book. Hence, for multivariate normal papers this bibliography plus Anderson's textbook will cover papers through 1972.

The preparation of a bibliography can be compared to Samuel Johnson's statement about dictionaries: "No dictionary of a living tongue ever can be perfect, since while it is hastening to publication, some words are budding and some falling away."

D. B. Owen

PREFACE

This monograph is an outcome of the authors' attempts at collecting the numerous publications that have appeared in the recent years. In view of the interrelatedness of the papers, it was decided to extend the project to cover the period since Anderson's book on Multivariate Analysis first appeared (1958). It was obvious to us that for the material to be useful, one would not only have to classify each article, but would also have to abstract it - bringing out its most important features. The present book is the outcome of several hundred hours of joint effort.

Since the modus operandi involved the reading of each of the papers quoted, we had the help of several people: Authors who were kind enough to send very promptly their pre-prints or re-prints, staff of the University of Manitoba's E. Dafoe Library, the inter-library facilities of the University of Manitoba. A special word of gratitude to Mrs. Eva Loewen for the careful typing of the manuscript several times to suit our demands on style and spacing. The University of Manitoba provided the time by granting a sabbatical to the first author. Last, but not the least, our sons, Rama and Narayana, added their mite by providing us with their patient appreciation of the immensity of the task and kept their demands on our time to a minimum.

But for Dr. D. B. Owen, who kindly brought it to the attention of the publishers, this book would never have seen the light of day.

We are thankful to him for this and for agreeing to write a foreword. Financial help was rendered to both authors by the National Research Council of Canada. We gratefully acknowledge its role in the preparation of the manuscript.

This work is thus a joint effort of a number of helpful people.

Winnipeg | Kocherlakota Subrahmaniam
July 1973 | Kathleen Subrahmaniam

INTRODUCTION

In the recent years there has been an immense spurt of research in multivariate analysis which has not been documented collectively. This sudden interest and the resulting spate of publications was led to a large extent by the computers and the demands of other disciplines on the statistician. Although a number of books have appeared since Anderson (1) published his inspiring text, none seems to have incorporated the latest contributions. A recent bibliography (2) of Anderson-Das Gupta-Styan has alleviated this to some extent. Unfortunately, their book, aside from being slightly out-dated, has the additional disadvantage of including in it all the papers involving 'several' variables. Also, a mere classification of the articles, to our minds, does not help a researcher to any great extent. To this end, we decided to restrict our scope to multivariate normal and related distributions, adding depth to the study by abstracting each of the selected papers. As most of the work of greater interest and impact has appeared since 1957, and as Anderson (1) has an excellent bibliography of all previous work, it was decided to include papers published in the period 1957-1972.

The method of operation involved the collection of copies of all the articles and reading each thoroughly, with a view to presenting its main features as briefly as possible. No attempt has been made to critically review the article. This would have seriously undermined the effectiveness and the usefulness of the monograph. The only subjective part of the book lies in the choice of the topics; they had to deal with analysis based on the multivariate normal distribution. Some natural selection may have resulted from the obscurity of a publication.

The use of the bibliography is quite straightforward. A prospective reader can either select a topic and look in the subject index and survey the papers published on this field through the abstracts, or he may directly locate the articles of interest through the author index. Each author has been catalogued chronologically. If more than one author is involved, the ordering is in accordance with the initial letters of the last names. Single authors always enjoy a precedence in the same year. Some of the articles which came to our attention too late or for which we were unable to obtain a copy are listed separately. Fortunately, this is a very short list. The style of abbreviating the journals is similar to the one adopted by the American Mathematical Society. For consideration of space we have tried to make the abstract as brief as possible. This may have resulted in a terseness of style. We have to apologize for this.

A word as to the abbreviations used in the text would be in order. All abbreviations should be clear from the context. The most commonly occurring ones are: MV (multivariate), BVN/TVN (bi-/tri-variate normal), LR (likelihood ratio) and SCI/SCB (simultaneous confidence intervals/bounds).

No books have been abstracted in this monograph and as such no list of books is appended. We refer to (2) for a detailed and annotated listing of texts. The few books that have appeared in print since then and which we feel may be relevant are listed below:

(1) Anderson, T.W. (1958): An Introduction to Multivariate Statistical

Analysis. John Wiley & Sons, New York.

(2) Anderson, T.W., Das Gupta, S. and Styan, G.P.H. (1972): *A Bibliography of Multivariate Statistical Analysis*. A Halsted Press Book, John Wiley & Sons, New York.

(3) Dempster, A.P. (1969): *Elements of Continuous Multivariate Analysis*. Addison-Wesley Publishing Company, Reading, Mass.

(4) Johnson, N.L. and Kotz, S. (1972): *Distributions in Statistics: Continuous Multivariate Distributions*. John Wiley & Sons, New York.

(5) Kshirasagar, A.M. (1972): *Multivariate Analysis*. Marcel Dekker, Inc., New York.

(6) Morrison, D. (1967): *Multivariate Statistical Methods*. McGraw-Hill Book Company, New York.

(7) Press, S. James (1972): *Applied Multivariate Analysis*. Holt, Rinehart and Winston, Inc., New York.

(8) Rao, C.R. (1965): *Linear Statistical Inference and Its Applications*. John Wiley & Sons, New York.

(9) Roy, S.N., Gnanadesikan, R. and Srivastava, J.N. (1971): *Analysis and Design of Certain Quantitative Multiresponse Experiments*. Pergamon Press, New York.

MULTIVARIATE ANALYSIS

A Selected and Abstracted Bibliography, 1957-1972

1. Abrahamson, I.G. (1964): Orthant probabilities for the quadrivariate normal distribution. Ann. Math. Statist., 35, 1685-1703.

Standardized quadrivariate normal. Expression of probability in positive quadrant as the sum of orthoscheme probabilities. Special cases of bivariate and trivariate normal in closed form.

Adams, J.W. (1964): See Kotz, S. and Adams, J.W., #634.

2. Adke, S.R. (1958): A note on distance between two populations. Sankhya, 19, 195-200. [Correction: 20 (1959), 108].

Distance function and its interpretation. Jeffreys's invariants and D^2, T^2 statistics. Distributions admitting sufficient statistics.

3. Afifi, A.A. and Elashoff, R.M. (1966): Missing observations in multivariate statistics I. Review of the literature. J. Amer. Statist. Assoc., 61, 595-604.

A review of the literature on statistical analysis in the p-dimensional case with missing observations.

4. Afifi, A.A. and Elashoff, R.M. (1967): Missing observations in multivariate statistics II. Point estimation in simple linear regression. J. Amer. Statist. Assoc., 62, 10-29.

Missing data. Estimation of simple linear regression between y and x when (x,y) are general bivariate random variables. Least squares, zero order, modified zero order and first and second mixed methods of estimation.

5. Afifi, A.A. and Elashoff, R.M. (1969): Multivariate two sample tests with dichotomous and continuous variables I. The location model. Ann. Math. Statist., 40, 290-298.

Tests of hypotheses with continuous/dichotomous vectors. Development of T^2, information theoretic and L.R. tests. Mixtures of multinormals. Discriminant analysis for location model. Asymptotic distributions of statistics.

6. Afifi, A.A. and Elashoff, R.M. (1969): Missing observations in multivariate statistics III. Large sample analysis of simple linear regression. J. Amer. Statist. Assoc., 64, 337-358.

Estimation in general bivariate distributions. Linear regression. Least squares. Asymptotic properties of the estimators.

7. Afifi, A.A. and Elashoff, R.M. (1969): Missing observations in multivariate statistics IV. A note on simple linear regression. J. Amer. Statist. Assoc., 64, 359-365.

Randomly occurring missing observations. Small sample properties of estimators in simple linear regression for general bivariate distributions. Monte Carlo study.

Afifi, A.A. (1972): See Azen, S.P. and Afifi, A.A., #54.

8. Afonja, B. (1972): The moments of the maximum of correlated normal and t-variates. J. Roy. Statist. Soc., B, 34, 251-262.

Expressions for moments of $\max(x_1, \ldots, x_k)$ *in a k-dimensional distribution in terms of corresponding truncated standard distributions. Multivariate normal. Multivariate t when means are all equal.*

9. Ahamad, B. (1967): An analysis of crimes by the method of principal components. Appl. Statist., 16, 17-35.

Application of principal component analysis to crime data of England and Wales. Interpretation of first principal component as due to population change.

10. Ahamad, B. (1968): A note on the interrelationships of crimes - a reply to Mrs. Walker. Appl. Statist., 17, 49-51.

An answer to paper by Mrs. Walker. Defence of the above paper #9.

11. Ahsanullah, M. (1971): On the estimation of means in a bivariate normal distribution with equal marginal variances. Biometrika, 58, 230-233.

Estimation of one of the means in bivariate normal when variances are equal but unknown. Preliminary test of significance on the equality of the two means. Bias and MSE. Comparison of the usual estimator with PTS estimator. Determination of efficiencies. Tables of bias.

12. Aitchison, J. (1965): Likelihood ratio and confidence-region tests. J. Roy. Statist. Soc., B, 27, 245-250.

Confidence region tests for multiple hypotheses in the multivariate case. Relationship between confidence region tests and LR tests. Example of two multivariate normals. Tests for difference of means by nested hypotheses on subvectors of the means.

13. Aitkin. M.A. (1964): Correlation in a singly truncated bivariate normal distribution. Psychometrika, 29, 263-270.

Truncated BVN with truncation in one variable. Expressions for moments in terms of moments of the complete distribution. Charts and tables for correlation.

14. Aitkin, M.A. and Hume, M.W. (1965): Correlation in a singly truncated bivariate normal distribution II. Rank correlation. Biometrika, 52, 639-643.

Expected values of Spearman's ρ and Kendall's τ for singly truncated BVN. Tables.

15. Aitkin, M.A. (1966): The correlation between variate-values and ranks in a doubly truncated normal distribution. Biometrika, 53, 281-282.

Correlation coefficient of variate-values and ranks in doubly truncated BVN. Table.

16. Aitkin, M.A. and Hume, M.W. (1966): Correlation in a singly truncated bivariate normal distribution III. Correlation between ranks and variate-values. Biometrika, 53, 278-281.

BVN distribution. Correlation between ranks and variate-values for the marginal distribution in singly truncated case. Table and chart.

17. Aitkin, M.A. and Hume, M.W. (1968): Correlation in a singly truncated bivariate normal distribution IV. Empirical variances of rank correlations coefficients. Biometrika, 55, 437-438.

Monte Carlo study. Variances of Kendall's τ and Spearman's ρ. Charts.

18. Aitkin, M.A., Nelson, W.C. and Reinfurt, K.H. (1968): Tests for correlation matrices. Biometrika, 55, 327-334.

Modified LR tests for hypotheses concerning correlation matrix P: (i) H_1: $P = \rho I$, (ii) H_2: $P = P_0$ (unspecified), and (iii) for trivariate case H_3: $\rho_{12} = \rho_{13}$ (unspecified). Asymptotic distributions of criteria in some cases. Heuristic justification.

19. Aitkin, M.A. (1969): Some tests for correlation matrices. Biometrika, 56, 443-446. [Correction: 58 (1971), 245].

Asymptotic distributions of statistics (i) T_1 for H_0: $P = P_0$ in one sample case, (ii) T_2 for testing H_0: $P_1 = P_2$ in the two sample case. Special cases of $p = 2, 3$. Simultaneous confidence bounds on ρ_{jk}. Union-Intersection principle.

20. Akaike, H. (1958): On a computation method for eigenvalue problems and its application to statistical analysis. Ann. Inst. Statist. Math., 10. 1-20.

Computational aspects of solution of eigenvalue problem $A x = \lambda Bx$. Classification into three or more groups. Canonical analysis.

21. Alam, K. and Rizvi, M.H. (1966): Selection from multivariate normal populations. Ann. Inst. Statist. Math., 18, 307-318.

Reduction of problem of ranking of k MVN populations in terms of Mahalanobis' distances to one of ranking in terms of noncentral χ^2 or F. Selection of t-largest of k populations and a subset containing the t largest.

Al-Ani, S. (1969): See Pillai, K.C.S., Al-Ani, S. and Jouris, G.M., #895.

22. Al-Ani, S. (1970): On the noncentral distributions of the second largest roots of three matrices in multivariate analysis. Canad. Math. Bull., 13, 299-304.

Distribution of the second largest latent root in three (non-null) cases: MANOVA, canonical correlations, tests for equality of covariance matrices. Limiting forms.

Al-Ani, S. (1970): See Pillai, K.C.S. and Al-Ani, S., #897.

23. Al-Ani, S. (1972): On the ith latent root of a complex matrix. Canad. Math. Bull., 15, 323-327.

Distribution of the second largest root of a complex matrix. Use to establish the distribution of the ith largest root. Central case.

Allen, D.M. (1969): See Grizzle, J.E. and Allen, D.M., #402.

24. Ames, E. and Reiter, S. (1961): Distributions of correlation coefficients in economic time series. J. Amer. Statist. Assoc., 56, 637-646.

Results of sampling experiment of 100 time series over 25 years. Coefficients of correlation, autocorrelation and their logarithms, with and without correction for trend.

25. Amos, D.E. (1969): On the computation of the bivariate normal distribution. Math. Comput., 23, 655-659.

Quadrature and series representations for BVN as limiting forms of BV t distribution. Bessel function and incomplete Beta series.

26. Amos, D.E. and Bulgren, W.G. (1969): On the computation of a bivariate t-distribution. Math. Comput., 23, 319-333.

Integral and series representations for the distribution function of a bivariate t. Computational procedures using special functions. Numerical evaluations.

27. Anderson, G.A. (1965): An asymptotic expansion for the distribution of the latent roots of the estimated covariance matrix. Ann. Math. Statist., 36, 1153-1173.

Expansion of $\int \exp[(-n/2) tr\ AHLH'](H'dH)$ as a series involving $1/n$ and the roots of the matrices A and L under assumption of strictly increasing population roots. Erdelyi's asymptotic procedures. Correlations for MLE when no information available and under uniform priors. Tables.

28. Anderson, G.A. (1970): An asymptotic expansion for the non-central Wishart distribution. Ann. Math. Statist., 41, 1700-1707.

Approximations and expansions for the distribution of non-central Wishart by (i) generalization of Laplace's method, (ii) James's differential equation method. Partial differential equation satisfied by $\int \exp tr(H'A)(H'dH)$. Bessel function of matrix argument.

29. Anderson, J.A. (1969): Constrained discrimination between k populations. J. Roy. Statist. Soc., B, 31, 123-139.

Problem of classification into k populations. Complete solution. Allocation rule maximizing probability of correct classification while satisfying upper bounds on those of misclassification.

30. Anderson, M.R. (1971): A characterization of the multivariate normal distribution. Ann. Math. Statist., 42, 824-827.

Characterizations of MVN by mean vector and variance matrix. Exponential form of the p.d.f.

31. Anderson, T.W. (1957): Maximum likelihood estimates for a multivariate normal distribution when some observations are missing. J. Amer. Statist. Assoc., 52, 200-203.

Missing data in MVN. With incomplete observation vectors. Estimation of mean and variance using conditional likelihood and regression techniques.

32. Anderson, T.W. and Bahadur, R.R. (1962): Classification into two multivariate normal distributions with different covariance matrices. Ann. Math. Statist., 33, 420-431.

Discrimination and tests of hypotheses for two sample case. $\Sigma_1 \neq \Sigma_2$. *Linear procedures for classification. Bayes and minimax procedures. Optimality.*

33. Anderson, T.W. (1963): Asymptotic theory for principal component analysis. Ann. Math. Statist., 34, 122-148.

Asymptotic distributions of roots and principal components based on covariance matrix and correlation matrix. Tests and estimation concerning roots and vectors. Multiple roots case. Discussion.

34. Anderson, T.W. (1963): A test for equaltiy of means when covariance matrices are unequal. Ann. Math. Statist., 34, 671-672.

Extension of Bennett-Scheffe solution to more than two groups.

35. Anderson, T.W. (1963): The use of factor analysis in the statistical analysis of multiple time series. Psychometrika, 28, 1-25.

Review paper discussing the roles of factor analysis, principal component analysis, canonical analysis, and regression in dealing with MV data. Examples.

36. Anderson, T.W. and Das Gupta, S. (1963): Some inequalities on characteristic roots of matrices. Biometrika, 50, 522-524. [Correction: 52 (1965), 669].

Inequalities generalizing Roy's inequalities for a positive definite matrix and functions of such a matrix, in terms of its roots. Case of two matrices.

37. Anderson, T.W. and Das Gupta, S. (1964): Monotonicity of the power functions and some tests of independence between two sets of variates. Ann. Math. Statist., 35, 206-208.

Class of procedures invariant under groups of transformations depending only on sample canonical correlations. Sufficient conditions for monotonicity of power function with respect to canonical correlations. LR and Roy's maximum root procedures and monotonicity of power.

38. Anderson, T.W. and Das Gupta, S. (1964): A monotonicity property of the power functions of some tests of the equality of two covariance matrices. Ann. Math. Statist., 35, 1059-1063. [Correction: 36 (1965), 1318].

Invariant tests based on roots of $S_1S_2^{-1}$ *with the power depending only on* $\Sigma_1\Sigma_2^{-1}$. *Sufficient conditions for an acceptance region to have monotically increasing power function. Examples. Tests for* H_0: $\Sigma = I$.

Anderson, T.W. (1964): See Das Gupta, S., Anderson, T.W. and Mudholkar, G.S., #217.

39. Anderson, T.W. (1965): Some optimum confidence bound for roots of determinantal equations. Ann. Math. Statist., 36, 468-488.

Simultaneous confidence bounds for (i) characteristic roots of Σ, *(ii) characteristic roots of* $\Sigma_1\Sigma_2^{-1}$, *(iii) some ordered roots in a two sample problem. Comparison with work of Roy and of Roy and Gnanadesikan. Tests of hypotheses for equality of covariance matrices.*

40. Anderson, T.W. (1965): Some properties of confidence regions and tests of parameters in multivariate distributions, (with discussion). Proc. IBM Sci. Comput. Symp. Statist., 15-28.

Review paper on inferences concerning $\Sigma_1 = \Sigma_2$. *An examination of several procedures and their properties. Confidence regions for* $\Sigma_1\Sigma_2^{-1}$. *Inequalities on the characteristic roots of* $\Sigma_1\Sigma_2^{-1}$.

41. Anderson, T.W. (1969): Statistical inference for covariance matrices with linear structure. Multivariate Anal.-II, [Proc. 2nd Internat. Symp., Krishnaiah, ed.], 55-66.

MLE for covariance matrix coefficients when both Σ *and its inverse are linear functions of known matrices. LR tests. Markov estimates. Computational aspects.*

42. Anderson, T.W. (1970): Estimation of covariance matrices which are linear combinations or whose inverses are linear combinations of given matrices. Essays Prob. Statist., (Roy Volume, Bose et al, eds.), 1-24.

MLE of covariance matrices which are (i) linear functions of known symmetric matrices, (ii) such that their inverses are linear functions of known symmetric matrices. Asymptotic properties. Tests of hypotheses concerning such models. Examples.

43. Andrews, D.F., Gnanadesikan, R. and Warner, J.L. (1971): Transformations of multivariate data. Biometrics, 27, 825-840.

Extension of Box-Cox procedures to multivariate case. Data-based power transformations resulting in either marginal or joint normality. Directional normality. Examples.

44. Andrews, D.F. (1972): Plots of high-dimensional data. Biometrics, 28, 125-136.

Data analytic techniques in multivariate analysis. Transformation of a point by $f(t) = x_1/\sqrt{2} + x_2 \sin t + x_3 \cos t + x_4 \sin 2t + x_5 \cos 2t + \ldots$ *Applications to discrimination, classification and cluster analysis.*

Andrews, H.P. (1971): See Snee, R.D. and Andrews, H.P., #1050.

45. Ando, A. and Kaufman, G.M. (1965): Bayesian analysis of the independent multinormal process - neither mean nor precision known. J. Amer. Statist. Assoc., 60, 347-358.

Natural conjugate family of prior densities for MVN. Prior posterior and pre-posterior analyses.

Antle, C.E. (1970): See Kappenman, R.F., Geisser, S. and Antle, C.E., #573.

46. Appleby, R.H. and Freund, R.J. (1962): An empirical evaluation of multivariate sequential procedure for testing means. Ann. Math. Statist., 33, 1413-1420.

Monte Carlo evaluation of sequential procedures based on χ^2 *(*Σ *known) and* T^2 *(*Σ *unknown). Tables of average sample numbers and of observed* α *and* β*.*

Armitage, J.V. (1965): See Krishnaiah, P.R. and Armitage, J.V., #647.

Armitage, J.V. (1966): See Krishnaiah, P.R. and Armitage, J.V., #649.

Armitage, J.V. (1969): See Krishnaiah, P.R., Armitage, J.V. and Breiter, M.C., #657.

Armitage, J.V. (1969): See Krishnaiah, P.R., Armitage, J.V. and Breiter, M.C., #658.

Armitage, J.V. (1970): See Krishnaiah, P.R. and Armitage, J.V., #659.

47. Arnold. B.C. (1967): A note on multivariate distributions with specified marginals. J. Amer. Statist. Assoc., 62, 1460-1461.

Generalization of Marshall and Olkin procedures for generating MV distributions with marginals in a prescribed class.

48. Arnold. B.C. (1968): Parameter estimation for a multivariate exponential distribution. J. Amer. Statist. Assoc., 63, 848-852.

Simple consistent estimates of the parameters. MSE. Lower bound for efficiency.

49. Arnold, H.J. (1964): Permutation support for multivariate techniques. Biometrika, 51, 65-70.

Robustness of Hotelling's T^2. *Permutation cumulants. BVN, rectangular and double exponential populations. Monte Carlo studies, comparison of actual and nominal levels of significance; adjustments.*

50. Asano, C. and Sato, S. (1962): A bivariate analogue of pooling of data. Bull. Math. Statist., 10, 39-59.

Estimation based on preliminary tests of significance: (i) mean when $\Sigma_1 = \Sigma_1 = \Sigma$ *known/unknown, (ii) variance matrices, (iii) generalized variances. Distributions of estimators. Bias and MSE.*

Ashford, J.R. (1969): See Sowden, R.R. and Ashford, J.R., #1056.

51. Ashton, E.H., Healy, M.J.R. and Lipton, S. (1957): The descriptive use of discriminant functions in physical anthropology. Proc. Roy. Soc. London Ser. B, 146, 552-572.

Multiple discriminant functions for more than two populations. Replacing the original p measurements by a set of k linear functions such that the sum of all the D^2 *between different pairs of populations is maximized. Technique illustrated by anthropological data.*

52. Asoh, Y.H. and Okamoto, M. (1969): A note on non-null distribution of the Wilks statistic in MANOVA. Ann. Inst. Statist. Math. 21, 67-71.

Expressing LR statistic as the product of conditional betas in the non-null case. Lower bound of Wilks statistic.

53. Asoo, Y. (1970): Note on the estimation of the standardized covariance matrix. Ann. Inst. Statist. Math., 22, 175-179.

Estimation of correlation coefficients when variances are known. Asymptotic relative efficiency of the usual estimator $\{r_{ij}\}$ *Example for* $p = 2, 3$.

54. Azen, S.P. and Afifi, A.A. (1972): Asymptotic and small-sample behaviour of estimated Bayes rules for classifying time-dependent observations. Biometrics, 28, 989-998.

Classification into two populations. Data collected over time with polynomial trend and autocorrelated errors. Bayes rule procedures. Comparison of estimators: (i) MLE of all parameters, (ii) ignoring autocorrelations, remaining estimation by least-squares. Monte Carlo study of estimators. Estimating probability of misclassification.

55. Bacon, R.H. (1963): Approximations to multivariate normal orthant probabilities. Ann. Math. Statist., 34, 191-198.

Approximations to $P\{X > 0\}$ *when X is MVN (standard) with correlation matrix P. Recursive relations for general dimensionality. Comparisons with previous results. Examples.*

56. Bagai, O.P. (1962): Statistics proposed for various tests of hypotheses and their distributions in particular cases. Sankhya, A, 24, 409-418. [Correction: 25 (1963), 427].

Null distributions of six statistics in multivariate analysis: Roy's largest root, T_0^2, Λ, U, *Pillai's V, and Bagai's Y. Exact distribution for* $p = 2, 3$. *Limiting forms for* $p = 2$ *to 6.*

57. Bagai, O.P. (1962): Distribution of the determinant of the sum of products matrix in the noncentral linear case for some values of p. Sankhya, A, 24, 55-62. [Correction: 25 (1963), 428].

The distribution of $|A|$ *in the noncentral linear case, i.e., when noncentrality matrix is of rank one. Special cases only.* $p = 2, 3, 4$ *are considered. Representation in special integrals.*

58. Bagai, O.P. (1964): Distribution of the ratio of the generalized variance to any of its principal minors. J. Indian Statist. Assoc., 2, 80-96.

Generalization of Wilks' results. Case of $p = 3(1)10$ *considered. Integral representation.*

59. Bagai, O.P. (1964): Limiting distribution of some statistics in multivariate analysis of variance. Sankhya, A, 26, 271-278.

Limiting distribution of Wilks-Lawley U and Bagai Y statistics for p = 5(1)10 in the null case.

60. Bagai, O.P. (1965): The distribution of the generalized variance. Ann. Math. Statist., 36, 120-130.

Extension of #57. Distribution of |A| in the noncentral linear case for p = 2(1)10. Integral representation.

61. Bagai, O.P. (1965): The distribution of some multivariate test statistics. J. Indian Statist. Assoc., 3, 116-124.

Tests of hypotheses on the covariance matrix. Exact null distribution of the sphericity test statistic for any p. Representation as a multiple series. Distribution of the U statistic.

62. Bagai, O.P. (1967): The distribution of the ratio of two generalized variances. J. Indian Statist. Assoc., 5, 124-138.

Exact null distribution of the ratio of two generalized (independent) variances for the case p = 2(1)7.

63. Bagai, O.P. (1969): Approximate distributions of non-orthogonal complex estimates in analysis of variance. J. Indian Statist. Assoc., 7, 82-90.

Multivariate analogue of Satterthwaites' technique. Approximations to the linear function of independent Wisharts by a single Wishart. Application to multivariate general linear model.

64. Bagai, O.P. (1972): On the exact distribution of some p-variate test statistics with the use of G-function. Sankhya, A, 34, 171-186.

Mellin transform techniques and Meijer's G-function representation. Exact null distribution of Wilks' Λ, sphericity test statistic, independence of sets of variates and ratio of generalized variances.

65. Bagai, O.P. (1972): On the exact distribution of a test statistic in multivariate analysis of variance. Sankhya, A, 34, 187-190.

Exact null distribution of Bagai's Y statistic. Tests for equality of means, equality of covariance matrices and independence of sets of variates. Mellin transform techniques and hypergeometric function representation. p = 2(1)8.

66. Bahadur, R.R. (1961): On classification based on responses to n dichotomous items. Stud. Item. Anal. Predict., (Solomon, ed.), 169-176.

Classification and distance problems. Qualitative characters of dichotomous type. Definition of distance and the problem of classification based on LR. Asymptotic normality of criterion. Approximations to LR statistic and the distance function.

Bahadur, R.R. (1962): See Anderson, T.W. and Bahadur, R.R. , #32.

Bain, L.J. (1972): See Bemis, B.M., Bain, L.J. and Higgins, J.J. , #86.

67. Balakrishnan, V. and Sanghvi, L.D. (1968): Distance between populations on the basis of attribute data. Biometrics, 24, 859-865.

Measurement of distance between populations based on attribute data. Definition of a new index. Comparison with (i) Edwards and Cavalli-Sforza (ii) Steinberg indices. Geometrical interpretation.

Bancroft, T.A. (1967): See Srivastava, S.R. and Bancroft, T.A. , #1080.

Bandyopadhyay, S. (1964): See Mukherjee, R. and Bandyopadhyay, S. , #823.

68. Banerjee, D.P. (1958): On the exact distribution of a test in multivariate analysis. J. Roy. Statist. Soc., B, 20, 108-110.

LR tests on means. Distribution of the statistic L by Mellin transform method.

69. Banerjee, D.P. (1960): On the forms of some invariants of probability distribution. Metron, 20, 299-306.

Jeffreys invariants for distance between populations. Koopman's class, Beta, Pearson's type III distributions.

70. Banerjee, K.S. and Marcus, L.F. (1968): Bounds in a minimax classification procedure. Biometrika, 55, 653-654.

Anderson-Bahadur classification procedure when $\Sigma_1 \neq \Sigma_2$. Bounds on a parameter t.

71. Banerjee, S. (1962): The problem of testing linear hypothesis about population means when the population variances are not equal and M-test. Sankhya, A, 24, 363-376.

Tests on linear functions of means from k MVN populations ignoring covariances.

Bantegui, C.G. (1959): See Pillai, K.C.S. and Bantegui, C.G., #876.

72. Baranchik, A.J. (1970): A family of minimax estimators of the mean of a multivariate normal distribution. Ann. Math. Statist., 41, 642-645.

A minimax procedure which dominates the usual estimators for quadratic loss function with variance matrix known.

Bargmann, R.E. (1958): See Roy, S.N. and Bargmann, R.E., #966.

73. Bargmann, R.E. (1961): Multivariate statistical analysis in psychology and education. Bull. Inst. Internat. Statist., 38, 79-86.

Uses of Hotelling's T^2, MANOVA, factor analysis, canonical correlation analysis in psychology. Examples.

Bargmann, R.E. (1964): See Posten, H.O. and Bargmann, R.E., #908.

Bargmann, R.E. (1964): See Trawinski, I.M. and Bargmann, R.E., #1133.

74. Bargmann, R.E. (1969): Exploratory techniques involving artificial variables. Multivariate Anal.-II, [Proc. 2nd Internat.Symp., Krishnaiah, ed.], 567-580.

Pitfalls in common MV techniques and their interpretations: (i) discriminant analysis, (ii) first principal component, (iii) model building using latent class analysis, (iv) factor analysis. Examples given.

75. Bargmann, R.E. (1970): Interpretation and use of a generalized discriminant function. Essays Prob. Statist., (Roy Volume, Bose et al, eds.), 35-60.

Interpretation of linear combinations of response variables as generalized discriminant functions. Noncentrality parameters. Illustrations: (i) two groups, (ii) k groups, (iii) two way classification, (iv) multiple regression. Use in variable selection.

76. Barnett, V.D. and Lewis, T. (1963): A study of the relation between G.C.E. and degree results. J. Roy. Statist. Soc., A, 126, 178-226.

Canonical analysis for degree result prediction using age, sex, type of schooling as linear predictors. Contrasting between canonical analysis and multiple regression. Extensive data. Discussion.

77. Barten, A.P. (1962): Note on unbiased estimation of the squared multiple correlation coefficient. Statistica Neerlandica, 16, 151-163.

Multiple correlation coefficient in the conditional case. Expectation to $O(n^{-2})$ and correction of bias. Derivation of unbiased estimator. Tables.

78. Bartlett, M.S. and Please, N.W. (1963): Discrimination in the case of zero mean differences. Biometrika, 50, 17-21.

Discrimination when means are equal and variances are unequal. Quadratic function of the observations. Special variance-matrix form.

79. Bartlett, M.S. (1965): Multivariate statistics. Theor. Math. Biol., (Waterman and Morowitz, eds.), 201-224.

General and expository paper. Discusses: discrimination, size-shape factors, zero mean differences, more than two groups, canonical analysis.

80. Barton, D.E. and David, F.N. (1962): Randomization bases for multivariate tests. I. The bivariate case. Randomness of n points in a plane. Bull. Inst. Internat. Statist., 39, 455-467.

Cluster analysis of n sets of points. Random placement of points and tests for randomness. Expectation and variance of the test statistic. Approximation to Beta distribution. Nonparametric technique.

81. Basu, A.P. (1969): On some tests for several linear relations. J. Roy. Statist. Soc., B, 31, 65-71.

Multivariate linear regression model. Tests for k linear relations among p variables. Wishart distribution and the distribution of a test statistic based on Wishart matrix. Test statistic which is sum of dependent T^2 statistics. Approximations by χ^2.

82. Beale, E.M.L., Kendall, M.G. and Mann, D.W. (1967): The discarding of variables in multivariate analysis. Biometrika, 54, 357-366.

Selection of optimum set of variables for (i) regression analysis, (ii) data reduction in MV analysis. Maximum multiple correlation and minimum multiple correlation techniques. Comparison with stepwise regression analysis and principal component analysis.

83. Beattie, A.W. (1962): Truncation in two variates to maximize a function of the means of a normal multivariate distribution. Austral. J. Statist., 4, 1-3.

Truncation of a standard MVN distribution with known correlation along two co-ordinates so that for a given proportion retained the linear compound of the mean is maximum. Application to genetics.

84. Behara, M. and Giri, N. (1971): Locally and asymptotically minimax tests of some multivariate decision problems. Arch. Math., 22, 436-441.

Tests on mean vectors and subvectors when Σ not known. Minimax and locally, asymptotically minimax procedures. Invariant tests.

85. Bell, C.B. and Haller, H.S. (1969): Bivariate symmetry tests: Parametric and nonparametric. Ann. Math. Statist., 40, 259-269.

Definition of symmetry in BV case. LR tests for BVN and distribution of statistics. Comparison with non-parametric procedures.

86. Bemis, B.M., Bain, L.J. and Higgins, J.J. (1972): Estimation and hypothesis testing for the parameters of a bivariate exponential distribution. J. Amer. Statist. Assoc., 67, 927-929.

Comparison of ML and MM estimation. Usual sample correlation coefficient is unsatisfactory estimator for ρ. Test for $\rho = 0$ (independence of X and Y). Power of the test.

87. Bennett, B.M. (1957): Tests for linearity of regression involving correlated observations. Ann. Inst. Statist. Math., 8, 193-195.

Repeated measurements on same subject. Linearity of regression of the responses when Y's are MVN. Canonical transformation. Test based on modified F-ratio.

88. Bennett, B.M. (1957): On the performance characteristic of certain methods of determining confidence limits. Sankhya, 18, 1-12.

Comparison of confidence interval and fudicial interval for E(Y)/E(X) when (X,Y) are BVN. Variances and correlation coefficient are assumed known.

89. Bennett, B.M. (1959): On a multivariate version of Fieller's theorem. J. Roy. Statist. Soc., B, 21, 59-62.

Generalization of Fieller's theorem for p = 2. Confidence interval for ratio of means of two MVN populations when the variance matrices are unknown and unequal.

90. Bennett, B.M. (1961): Confidence limits for multivariate ratios. J. Roy. Statist. Soc., B, 23, 108-112.

Extension of #89. Confidence intervals for (i) ratio of means in a bivariate situation, (ii) ratio of regression coefficients for simple linear regression, (iii) ratio of means in successive multivariate samples. Use of Hotelling's T^2 and Wilks' Λ.

91. Bennett, B.M. (1961): On a certain multivariate nonnormal distribution. Proc. Cambridge Philos. Soc., 57, 434-436.

Multivariate t distribution. Estimation of parameters and their properties. Regression properties.

92. Bennett, B.M. (1963): On combining estimates of a ratio of means. J. Roy. Statist. Soc., B, 25, 201-205.

Combining independent estimates for ratio of means in a BVN with variance matrix equal to $\sigma^2 P$; P known and σ^2 unknown. Confidence intervals.

93. Bennett, B.M. (1964): A note on combining correlated estimates of a ratio of multivariate means. Technometrics, 6, 463-467.

Continuation of #92. Combining correlated estimates of ratio of means of two MVN samples. Confidence intervals based on Hotelling's T^2. Weighted least squares estimation.

94. Bennett, B.M. (1966): Multivariate "Analysis of Dispersion" in the presence of intra-class correlation. Metrika, 10, 1-5.

On some statistics having Wishart distribution in the presence of familial correlations. Equal covariance case. Application to analysis of dispersion and a modification.

95. Bennett, B.M. (1966): On alternatives to the multinormal distribution. Trabajos Estadist., 17, 45-51.

Two alternatives to MVN: Beta and MV t. Linearity of regression and marginal distributions. ML estimation of the parameters. Distribution of quadratic forms.

96. Bennett, G.W. and Cornish, E.A. (1963): A comparison of the simultaneous fiducial distributions derived from the multivariate normal distribution. Bull. Inst. Internat. Statist., 40, 902-919.

Simultaneous fiducial distribution of several parameters in MVN. Comparisons with (i) Segal, (ii) Fisher, (iii) marginals of the means, (iv) Cornish for BVN with equal variances. Estimation of pivotal quantities.

97. Bennett, G.W. (1964): The fiducial distribution of the parameters of a p-variate normal distribution with variance-covariance matrix $\Sigma = \sigma^2(1-\rho)I + \sigma^2\rho 11'$. CSIRO, Div. Math. Statist., Tech. Paper #18, 1-8.

Joint fiducial distribution of mean vector, ρ, σ^2 from a MVN, when $\Sigma = \sigma^2(1-\rho)I + \sigma^2\rho 11'$.

98. Berger, T. (1972): On the correlation coefficient of a bivariate, equal variance, complex Gaussian sample. Ann. Math. Statist., 43, 2000-2003.

X and A are complex BVN and bivariate Wishart respectively. Distribution of complex correlation coefficient $u = 2A_{12}/(A_{11}+A_{22})$. Expectation of $|u|^k$ and its behaviour for $n \to \infty$. Shows limit in the rth mean of $|u|$ is $|\rho|$ for all $r > 0$. Applications to spectral analysis.

99. Berk, R.H. (1969): Biorthogonal and dual configurations and the reciprocal normal distribution. Ann. Math. Statist., 40, 393-398.

Reciprocal normal distribution. Relationship of biorthogonal and dual configurations. Spherical distributions (invariance under orthogonal transformations). Wishart distribution. Partitioned vectors. Gauss-Markov theorem. Applications to distribution theory of quadratic forms.

100. Bhapkar, V.P. (1959): A note on multiple independence under multivariate normal linear models. Ann. Math. Statist., 30, 1248-1251.

Use of Roy's Union-Intersection principle for testing independence: Simultaneous confidence bounds on covariances. MV linear models. Roy's stepdown procedure. Examples.

101. Bhapkar, V.P. (1960): Confidence bounds connected with ANOVA and MANOVA for balanced and partially balanced incomplete block designs. Ann. Math. Statist., 31, 741-748.

Application of Roy's UI principle for setting up simultaneous confidence bounds on the total and partial parametric functions in MANOVA. Explicit algebraic expressions for ANOVA, MANOVA, BIBD and PBIBD.

102. Bhargava, R.P. (1966): Estimation of correlation coefficient when no simultaneous observations on the variables are available. Sankhya, A, 28, 1-14.

Point estimation of ρ with n units of $X_1 > x_0$ and X_2 measured on them. Two cases (i) X_1 is normal with mean and variance, (ii) X_1 is exponential. Loss function is the number of units required to sample for n of $X_1 > x_0$. Determination of x_0 so as to minimize expected loss under conditions (i) $\rho^2 \geq \rho_0^2$, ρ_0^2 known, (ii) variance of estimator less than or equal to D_0, pre-assigned. MLE for BVN model. Efficiency, Tables.

103. Bhargava, R.P. (1971): A test for equality of means of multivariate normal distributions when covariance matrices are unequal. Calcutta Statist. Assoc. Bull., 20, 153-156.

Testing quality of k means when variance matrices are unknown and unequal. LR procedure by nesting. Comparison with Anderson's technique. Asymptotic distribution of criterion.

104. Bhattacharyya, G.K. (1967): Asymptotic efficiency of multivariate normal score test. Ann. Math. Statist., 38, 1753-1758.

Nonparametric approach to testing hypothesis of shift in location of two populations. Comparison with Hotelling's T^2, MV analog of Wilcoxon's test and MV version of normal score test.

105. Bhattacharyya, G.K. and Johnson, R.A. (1968): Approach to degeneracy and the efficiency of some multivariate tests. Ann. Math. Statist., 39, 1654-1660.

Asymptotic relative efficiency of Hotelling's T^2, MV extension of Wilcoxon's test, and normal score test. Tests for shift in location. Corrects Bickel's ideas.

106. Bhattacharyya, G.K., Johnson, R.A. and Neave, H.R. (1971): A comparative power study of the bivariate rank sum test and T^2. Technometrics, 13, 191-198.

Small sample performance of bivariate two sample Wilcoxon's test. Shifts in one co-ordinate or both co-ordinates. Power comparison with T^2. Monte Carlo study.

107. Bhattacharya, P.K. (1966): Estimating the mean of a multivariate normal population with general quadratic loss function. Ann. Math. Statist., 37, 1819-1824.

Estimation of the mean vector with quadratic loss function. Upper bound on the risk function of this estimator. Comparison with usual estimator. Application to least squares in linear model.

108. Bildikar, S. and Patil, G.P. (1968): Multivariate exponential-type distributions. Ann. Math. Statist., 39, 1316-1326.

Characterizations of MV exponential type distributions including MVN. Characterizations based on cumulants and regressions.

109. Bildikar, S. (1969): On characterization of certain exponential-type distributions. Sankhya, B, 31, 35-42.

Characterization of MVN in terms of cumulants.

110. Bingham, C. (1972): An asymptotic expansion for the distribution of the eigenvalues of a 3x3 Wishart matrix. Ann. Math. Statist. 43, 1498-1506.

Distribution represented as the sum of products of confluent hypergeometric functions. Approximation through terms in n^{-8}. First term (product of Bessel functions) sufficiently accurate for many applications.

111. Birnbaum, A. and Maxwell, A.E. (1960): Classification procedures based on Bayes's formula. Appl. Statist., 9, 152-169.

Classification procedures based on Bayes's rule with no assumptions about form of distribution. Construction of admissible inference functions and their interpretations. Sampling errors. Comparison with other methods. Numerical examples.

112. Birren, J.E. and Morrison, D.F. (1961): Analysis of the WAIS subtests in relation to age and education. J. Gerontology, 16, 363-369.

Correlations (total and partial), principal component analysis, multiple regression and canonical correlation are illustrated. Data consist of age, education and components of IQ tests. Interpretation of principal components.

113. Blackith, R.E. (1960): A symthesis of multivariate techniques to distinguish patterns of growth in grasshoppers. Biometrics, 16, 28-40.

Describing growth in biological systems. Relative merits and relationships among various MV techniques: (i) discriminant analysis, (ii) factor analysis, (iii) principal component analysis, (iv) canonical analysis.

114. Bland, R.P. and Owen, D.B. (1966): A note on singular normal distributions. Ann. Inst. Statist. Math., 18, 113-116.

Standardized equicorrelated MVN when $|\Sigma| = 0$ $(\rho_{ij} = -1/(n-1)$ *for* $i \neq j)$. *Evaluation of* $Pr\{X \leq t\}$.

115. Bock, R.D. (1963): Programming univariate and multivariate analysis of variance. Technometrics, 5, 95-117.

Computational aspects of MANOVA: (i) canonical analysis, (ii) stepdown procedure. Relationship of latter with univariate ANOVA and analysis of covariance. MV analogue of analysis of covariance.

116. Bofinger, E. and Bofinger, V.J. (1965): The correlation of maxima in samples drawn from a bivariate normal distribution. Austral. J. Statist., 7, 57-61.

Standard BVN. Correlation of X_{max}, Y_{max} *for independent observations. Series expansion around* $\rho = 0$.

Bofinger, V.J. (1965): See Bofinger, E. and Bofinger, V.J., #116.

117. Bofinger, V.J. (1970): The correlation of maxima in several bivariate non-normal distributions. Austral. J. Statist., 12, 1-7.

Correlation of X_{max}, Y_{max} *when* (X,Y) *are from bivariate nonnormal distributions: Gamma, uniform, Poisson and binomial.*

118. Bolshev, L.N. (1969): Cluster analysis. Bull. Inst. Internat. Statist., 43, 411-425.

Review of literature from the Russian school.

119. Bosch, A.J. (1972): From univariate- to multivariate- analysis. Statistica Neerlandica, 26, 15-20.

A list of analogies between univariate and MVN. Several examples and introductory results.

Boswell, M.T. (1970): See Patil, G.P. and Boswell, M.T., #865.

120. Bowker, A.H. (1960): A representation of Hotelling's T^2 and Anderson's classificaton statistic in terms of simple statistics. Contrib. Prob. Statist., (Hotelling Volume, Olkin et al, eds.), 142-149.

On a property of the Wishart distribution under random orthogonal transformations. Partitioned Wishart matrix. Representation of Hotelling's T^2 as a ratio of two χ^2's. Expression of Anderson's classificaton statistic in terms of elements of n independent 2x2 Wisharts. Non-null cases.

121. Bowker, A.H. and Sitgreaves, R. (1961): An asymptotic expansion for the distribution function of the W-classification statistic. Stud. Item. Anal. Predict., (Solomon, ed.), 293-310.

Application of Bowker's results (#120). Expansion of the distribution function of W statistic. Characteristic functions technique.

122. Box, G.E.P. and Tiao, G.C. (1968): A Bayesian approach to some outlier problems. Biometrika, 55, 119-129.

Detection of outliers in the linear model using Bayesian procedures. Multivariate t distribution. Mixtures of normal distributions.

Boyce, R. (1962): See Chew, V. and Boyce, R., #161.

Bradley, R.A. (1961): See Jackson, J.E. and Bradley, R.A., #499.

Bradley, R.A. (1961): See Jackson, J.E. and Bradley, R.A., #500.

Bradley, R.A. (1966): See Jackson, J.E. and Bradley, R.A., #501.

Bradley, R.A. (1972): See Martin, D.C. and Bradley, R.A., #759.

Brandwood, L. (1959): See Cox, D.R. and Brandwood, L., #207.

Breiter, M.C. (1969): See Krishnaiah, P.R., Armitage, J.V. and Breiter, M.C., #657.

Breiter, M.C. (1969): See Krishnaiah, P.R., Armitage, J.V. and Breiter, M.C., #658.

123. Brillinger, D.R. (1963): Necessary and sufficient conditions for a statistical problem to be invariant under a Lie group. Ann. Math. Statist., 34, 492-500.

Group transformations. Example: cumulative distribution function of the sample correlation coefficient. Prior measure. Fiducial distribution.

124. Brindley, E.C., Jr. and Thompson, W.A., Jr. (1972): Dependence and aging aspects of multivariate survival. J. Amer. Statist. Assoc., 67, 822-830.

Review of literature on the MV exponential and related distributions.

125. Brogan, D.R. and Sedransk, J. (1971): Pooling correlated observations: Bayesian and preliminary tests of significance approaches. Tech. Rep. #280, Dept. Statist., U. Wisconsin.

Estimation in the BVN. Estimation of μ_Y using (i) preliminary test of significance on $\mu_Y = \mu_X$, (ii) regression estimator, (iii) prior distribution of μ_Y near μ_X. Comparison of preliminary test of significance and Bayesian approaches. Bias and MSE.

126. Brown, G.H. (1967): The use of correlated variables for preliminary culling. Biometrics, 23, 551-562.

Truncated BVN. Evaluation of BV integrals. Applications.

127. Brown, J.L. (1968): On the expansion of the bivariate Gaussian probability density using results of nonlinear theorems. IEEE Trans. Information Theory, 14, 158-159.

Expansion of BVN probability density function in terms of Hermite polynomials.

128. Buck, S.F. (1960): A method of estimation of missing values in multivariate data suitable for use with an electronic computer. J. Roy. Statist. Soc., B, 22, 302-306.

Estimation of missing values using regression techniques. Revision of variance matrix using estimated missing values. Bias associated with techniques. Comparison with ML procedures.

Bulgren, W.G. (1969): See Amos, D.E. and Bulgren, W.G., #26.

Bulgren, W.G. (1971): See Chase, G.R. and Bulgren, W.G., #148.

Bulgren, W.G. (1971): See Hewitt, J.E. and Bulgren, W.G., #461

129. Burnaby, T.P. (1966): Growth-invariant discriminant functions and generalized distances. Biometrics, 22, 96-110.

Partitioning of generalized D^2 into two additive components (taxonomically relevant and non-relevant factors). Comparison with principal component and canonical correlation analyses.

130. Bush, K.A. and Olkin, I. (1959): Extrema of quadratic forms with applications to statistics. Biometrika, 46, 483-486. [Correction: 48 (1961), 474-475].

Inequalities on quadratic forms (i) Min(WAW'), wB = d, (ii) max{(WAW')/(WBW')}. Applications to stratified sampling, moment problems, discrimination and canonical analysis.

131. Bush, K.A. and Olkin, I. (1961): Extrema of functions of a real symmetric matrix in terms of eigenvalues. Duke Math. J., 28, 143-152.

Inequalities on bilinear forms of the real symmetric matrix. Paley-matrix technique. Inequalities on quadratic forms. Eigenvalues.

132. Cacoullos, T. (1965): Comparing Mahalanobis distances I: Comparing distances between k known normal populations and another unknown. Sankhya, A, 27, 1-22.

Selection of population nearest to Π_0 out of k, with known common variance matrix. Constant loss function (i) natural indifference region: unique affine invariant minimax procedure, (ii) no natural indifference region: admissible minimax solution. Relation to discriminant analysis. Generalization of Mahalanobis distance.

133. Cacoullos, T. (1965): Comparing Mahalanobis distances II: Bayes procedures when the mean vectors are unknown. Sankhya, A, 27, 23-32.

Continuation of #132. Here (i) μ_0 and Σ known, (ii) only Σ known, (iii) no parameter known. Large sample criteria. Symmetric invariant Bayes procedures. Comparison of distances between pairs of populations.

134. Cacoullos, T. and Olkin, I. (1965): On the bias of functions of characteristic roots of a random matrix. Biometrika, 52, 87-94.

Inequalities satisfied by functions of matrix argument. Jensen's inequality in the matrix version. Roots of determinant equation. Functions of the roots of a matrix. Median bias of certain linear combinations of roots of an equation.

135. Cacoullos, T. (1966): On a class of admissible partitions. Ann. Math. Statist., 37, 189-195.

Partitioning of the space E_k of $\underline{X}$ into (k+1) convex (k dimensional) polyhedral cones of arbitrary probability content. Relationship to admissible partitions for family of classification and topothetical problems.

136. Cacoullos, T. (1967): Characterizations of normality by constant regression of linear statistics on another linear statistic. Ann. Math. Statist., 38, 1894-1898.

Characterization by the constant regression of a linear statistic on another linear statistic and by that of a set of linearly independent statistics on this statistic.

137. Cacoullos, T. (1967): Some characterizations of normality. Sankhya, A, 29, 399-404.

Characterizations of multinormality by constant regression of (i) a 'quadratic' function on the sum, (ii) a 'polynomial' on a linear function.

138. Castellan, N.J., Jr. (1966): On the estimation of the tetrachoric correlation coefficient. Psychometrika, 31, 67-73.

BVN. Rationale of tetrachoric correlation coefficient. Estimates and approximations. Comparison with Pearson's expressions.

139. Cattell, R.B. (1965): Factor analysis: an introduction to essentials I. The purpose and underlying models. Biometrics, 21, 190-215.

A review article outlining the models and purposes of factor analysis. Bibliographic paper.

140. Cattell, R.B. (1965): Factor analysis: an introduction to essentials II. The role of factor analysis in research. Biometrics, 21, 405-435.

Continuation of #139. Relations of factor analysis to other statistical techniques in scientific inference.

Cavalli-Sforza, L.L. (1965): See Edwards, A.W.F. and Cavalli-Sforza, L.L., #286.

141. Chaddha, R.L. and Marcus, L.F. (1968): An empirical comparison of distance statistics for populations with unequal covariance matrices. Biometrics, 24, 683-694.

Comparison of three statistics estimating distance squared between two multivariate normal populations with $\Sigma_1 \neq \Sigma_2$: *(i) Mahalanobis'* D^2, *(ii) Reyment's generalized distance, (iii) Anderson-Bahadur minimax criterion distance function. Comparisons based on two sets of data.*

142. Chakravarti, S. (1966): A note on multivariate analysis of variance test when dispersion matrices are different and unknown. Calcutta Statist. Assoc. Bull., 15, 75-86.

MANOVA when variance matrices are unequal and unknown. A new statistic based on jth sample variance matrix instead of pooled within variation. Approximation by linear function of χ^2*'s. Comparison with Ito-Schull work.*

143. Chakravarti, S. (1968): On the analysis of variance test in multivariate variance components model. Calcutta Statist. Assoc. Bull., 17, 57-78.

Test of hypothesis $\Sigma_i = 0$, $i = 1, 2, \ldots, k$, *in a variance-components model. Distribution of test statistic* $T = tr\, \Sigma(S_i S_0^{-1})$ *under the null and under the alternative. The joint distribution of the roots of* SS_0^{-1}. *MANOVA model II.*

144. Chambers, J.M. (1967): On methods of asymptotic approximations for multivariate distributions. Biometrika, 54, 367-383.

Multivariate approximations: Edgeworth expansions and the method of perturbations. Justification for Edgeworth type series. Examples: Wishart and noncentral Wishart. Perturbation methods: Approximate statistic rather than distribution. Examples: principal component analysis, canonical analysis, discriminant analysis.

145. Chambers, J.M. (1970): Computers in statistical research: simulation and computer-aided mathematics. Technometrics, 12, 1-15.

Simulation of probability models. Univariate and multivariate distributions. MVN, Wishart, noncentral Wishart.

146. Chan, L.S. and Dunn, O.J. (1972): The treatment of missing values in discriminant analysis-I. The sampling experiment. J. Amer. Statist. Assoc., 67, 473-477.

Monte Carlo studies. Probability of correct classification. Comparison of methods for handling missing values: (i) use only complete vectors, (ii) use all observations with no replacement, (iii) substitute means for missing values, (iv) Buck's regression method, (v) Dear's principal component method.

147. Chang, T.C. (1970): On an asymptotic representation of the distribution of the characteristic roots of $S_1S_2^{-1}$. Ann. Math. Statist., 41, 440-445.

Beta-type asymptotic representation of the distribution of the roots of $S_1S_2^{-1}$. Asymptotic expression for exact distribution as found by Khatri.

Chang, T.C. (1970): See Li, H.C., Pillai, K.C.S. and Chang, T.C., #725.

Chang, T.C. (1971): See Krishnaiah, P.R. and Chang, T.C., #661.

Chang, T.C. (1971): See Krishnaiah, P.R. and Chang, T.C., #662.

Chang, T.C. (1972): See Krishnaiah, P.R. and Chang, T.C., #665.

Chang, T.C. (1972): See Waikar, V.B., Chang, T.C. and Krishnaiah, P.R., #1151.

148. Chase, G.R. and Bulgren, W.G. (1971): A Monte Carlo investigation of the robustness of T^2. J. Amer. Statist. Assoc., 66, 499-502.

Behaviour of T^2 for 10, 5 and 1% levels of significance when distribution sampled is not normal: uniform, bivariate exponential, gamma, lognormal, double exponential. Samples of size 5, 10, 20. Tables of true and nominal α.

149. Chatterjee, S.K. (1959): On an extension of Stein's two sample procedure to the multi-normal problem. Calcutta Statist. Assoc. Bull., 8, 121-148.

First of a series. Sequential procedure for mean. Test whose power is free of nuisance parameters. Central and noncentral distributions in integral form. Monotonicity. Ellipsoidal regions. Case of $p = 2$. Choice of sample size. Kronecker's product of matrices - some new results.

150. Chatterjee, S.K. (1959): Some further results on the multinormal extension of Stein's two sample procedure. Calcutta Statist. Assoc. Bull., 9, 20-28.

Continuation of #149. Construction of tests and regions which are superior. Both depend on $\{\sigma_{ij}\}$. Less cost and more powerful.

151. Chatterjee, S.K. (1960): Sequential tests for the bivariate regression parameters with known power and related estimation procedures. Calcutta Statist. Assoc. Bull., 10, 19-34.

Inferences on regression coefficient β and intercept α in the conditional case of BVN. Sequential procedures and their termination with probability one. Distribution of sample size and the ASN.

152. Chatterjee, S.K. (1962): Sequential inference procedures of Stein's type for a class of multivariate regression problems. Ann. Math. Statist., 33, 1039-1064.

$E(y|x) = \alpha + \beta' x$, with x being multivariate normal. Sequential inference on β when $\sigma^2_{y|x}$ is unknown.

153. Chatterjee, S.K. (1962): Simultaneous confidence intervals of predetermined length based on sequential samples. Calcutta Statist. Assoc. Bull., 11, 144-149.

Construction of simultaneous confidence intervals for a set of (i) linear functions of means, (ii) linear function of regression coefficients in MVN case.

154. Chattopadhyay, A.K. and Pillai, K.C.S. (1970): On the maximization of an integral of a matrix function over the group of orthogonal matrices. Mimeo Series #248, Dept. Statist., Purdue U.

Inequalities involving characteristic roots. Maximization of zonal polynomials and hypergeometric function of matrix argument, with respect to $H \in O(p)$. Maximization of integral. Complex analogues.

155. Chattopadhyay, A.K. and Pillai, K.C.S. (1971): Asymptotic formulae for the distributions of some criteria for tests of equality of covariance matrices. J. Multivariate Anal., 1, 215-231.

Perturbation theory and Ito's methods for asymptotic distributions Expansions for distribution functions of (i) $m\ tr\ S_1S_2^{-1}$, (ii) $n\ U^{(p)}$, (iii) $F = \{m_1 tr(S_1S_4^{-1})/n_1 tr(S_3S_2^{-1})\}$. Max F. Tests on covariance matrices.

156. Chattopadhyay, A.K. and Pillai, K.C.S. (1971): Asymptotic expansions of the distributions of characteristic roots in MANOVA and canonical correlation. Mimeo Series #249, Dept. Statist., Purdue U.

Asymptotic expansions of the distribution functions of the characteristic roots in MANOVA and canonical correlation when (i) one extreme population multiple root, (ii) all distinct roots. Complex analogues.

157. Chen, C.W. (1971): On some problems in canonical correlation analysis. Biometrika, 58, 399-400.

Addition of variates in canonical correlation analysis. Canonical correlation between two sets of variates can never be decreased by addition of variates to either set.

158. Cheng, M.C. (1969): The orthant probabilities of four Gaussian variates. Ann. Math. Statist., 40, 152-161.

Expression of orthant probability of quadrivariate normal in terms of the inverse trigonometric functions, the dilogarithm function and its real part when correlation matrices are of particular forms.

159. Chentsov, N.N. (1967): On estimating an unknown mean of a multivariate normal distribution. Theory Prob. Appl., 12, 560-574.

Estimation of the mean vector given that $\Sigma = I$. Optimal estimator so that risk is minimized. Information inequality. Admissibility. Properties of information discriminant. Invariance.

160. Chernoff, H. (1972): Some measures for discriminating between normal multivariate distributions with equal covariance matrices. Tech. Rep. #73, Dept. Statist., Stanford U.

Classification procedures with unequal covariance matrices based on Anderson-Bahadur statistic and LR tests. Comparison with information approach. Minimizing maximum error probability.

Chetty, V.K. (1965): See Zellner, A. and Chetty, V.K., #1188.

161. Chew, V. and Boyce, R. (1962): Distribution of radial error in the bivariate elliptical normal distribution. Technometrics, 4, 138-140.

Density and distribution function of r, radial error, in a general BVN. Dependent and heteroscedastic errors. Tables for mean, median, mode and standard deviation of r/σ_y as function of σ_x/σ_y.

162. Chew, V. (1966): Confidence, prediction and tolerance regions for the multivariate normal distribution. J. Amer. Statist. Assoc., 61, 605-617.

Development of confidence, prediction and tolerance regions for the MVN for various cases of known and unknown mean vector and covariance matrix. Tables provided for BVN case. Bibliography.

163. Childs, D.R. (1967): Reduction of the multivariate normal integral to characteristic form. Biometrika, 54, 293-300.

Evaluation of the n-dimensional MVN integral over a generalized quadrant. Transformation to a series of integrals over entire hypervolume. Examples for n = 2, 3, 4 and for n = 6 and 9 with particular patterns in the correlation matrix.

164. Choi, S.C. (1971): Sequential test for correlation coefficients. J. Amer. Statist. Assoc., 66, 575-576.

Test based on sequences of sums and differences calculated at each step. Helmert transformation on sequences. Results in a sequential probability ratio test regarding parameter of Cauchy distribution. Table of average sample number compared with fixed sample size.

165. Choi, S.C. and Wette, R. (1972): A test for homogeneity of variances among correlated variables. Biometrics, 28, 589-592.

Testing equality of variances of several normal variables when covariances are unknown. Equality of covariances or correlation coefficients is assumed. Test based on multiple correlation coefficient of transformed variables.

Chou, C. (1971): See Siotani, M., Chou, C. and Geng, S., #1040.

166. Chou, C. and Siotani, M. (1972): Tables of upper 5 and 1% points of the ratio of two conditionally independent Hotelling's generalized T_0^2 statistics. Tech. Rep. #28, Dept. Statist., Kansas State U.

See #167 for discussion of theory. Tables given for $m_1 = 1(1)6, 8, 10$; $m_2 = 1(1)6, 8, 10$; $p = 2(1)6, 8, 10$; $n = 20(1)30(2)40, 45, 50, 60, 80, 120$. *Computer program for generating tables.*

167. Chou, C. and Siotani, M. (1972): Asymptotic expansion of the non-null distribution of the ratio of two conditionally independent Hotelling's T_0^2-statistics. Tech. Rep. #29, Dept. Statist., Kansas State U.

Ratio F_0 *of two Hotelling* T_0^2*-statistics. Tests of hypothesis* H_0: $M = 0$. *Conditional distribution for a given sample variance matrix. Perturbation theoretic methods. Asymptotic expansion of the distribution of* F_0 *in non-null case to* $O(n^{-2})$.

168. Choudhuri, S.B. (1960): Discriminant analysis of school final examination marks. Calcutta Statist. Assoc. Bull., 9, 123-138.

Tests on two groups of school results. Application of discriminant analysis. Classification.

169. Chow, G.C. (1966): A theorem on least squares and vector correlation in multivariate linear regression. J. Amer. Statist. Assoc., 61, 413-414.

Least squares estimate of the regression matrix maximizes the squared correlation between the matrix of observations and XB, i.e., $|Y'XB|^2/|Y'Y||B'X'XB|$.

170. Chung, J.H. and Fraser, D.A.S. (1958): Randomization tests for a multivariate two sample problem. J. Amer. Statist. Assoc., 53, 729-735.

Nonparametric alternatives to Hotelling's T^2 *to test for location parameters in two multivariate populations. Randomization techniques. Permutations. Large number of variables with too few observations.*

171. Church, A., Jr. (1966): Analysis of data when the response is a curve. Technometrics, 8, 229-246.

Application of principal component analysis to the data analysis of factorial designs. Interpretation of the vectors. Curvilinear responses. Analysis of residuals.

Clark, V. (1969): See Dunn, O.J. and Clark, V., #270.

Clark, V. (1971): See Dunn, O.J. and Clark, V., #272.

Clay, P.P.F. (1963): See Hopkins, J.W. and Clay, P.P.F., #475.

172. Cliff, N. (1970): The relation between sample and population characteristic vectors. Psychometrika, 35, 163-178.

Monte Carlo study of the sampling characteristics of the vectors of a reduced correlation matrix. Reduction achieved by substituting $R^2_{j.1,2,\ldots,p}$ *in place of 1 in the jth diagonal position.*

173. Clunies-Ross, C.W. and Riffenburgh, R.H. (1960): Geometry and linear discrimination. Biometrika, 47, 185-189.

Geometry of classificatory problem. Minimax principle. Best linear discriminator.

Cobb, W. (1960): See Roy, S.N. and Cobb, W., #974.

174. Cochran, W.G. and Hopkins, C.E. (1961): Some classification problems with multivariate qualitative data. Biometrics, 17, 10-32.

Classification into two or more groups when data are qualitative. Preliminary data to determine joint probability function. Consequences of constructing a rule from preliminary samples of finite size. Relative discriminating power of qualitative and continuous variates. Evaluation of probability of misclassification.

175. Cochran, W.G. (1964): Comparison of two methods of handling covariates in discriminatory analysis. Ann. Inst. Statist. Math., 16, 43-53.

Role of covariance analysis in discriminant and classificatory problems. Comparison of two techineques: (i) p discriminators and k covariates, (ii) (p+k) discriminators. Methods compared in terms of sample size needed to obtain equal power. Probability of misclassification when new specimens are classified.

176. Cochran, W.G. (1964): On the performance of the linear discriminant function. Technometrics, 6, 179-190.

Predicting probability of misclassification given by a discriminant function from probabilities given by variates used individually. Effect of positive and negative correlations. Examination of twelve sets of data from literature.

177. Cochran, W.G. (1968): Commentary on "Estimation of error rates in discriminant analysis". Technometrics, 10, 204-205.

Discussion of paper by Lachenbruch and Mickey (#701).

178. Cohen, A.C., Jr. (1957): Restriction and selection in multinormal distributions. Ann. Math. Statist., 28, 731-741.

MLE of parameters of MVN distribution when x_1 is restricted and remaining elements are unrestricted. Restrictions may be truncation, censoring or missing observations. Variance of the estimators. Applications.

179. Cohen, A. (1965): Estimates of linear combinations of the parameters in the mean vector of a multivariate distribution. Ann. Math. Statist., 36, 78-87.

Admissibility of $\gamma' y$ for $\phi'\theta$. All admissibly linear estimates. Essential unique Bayes solution. $C'y$ is admissible if and only if C lies in an ellipsoid. Inadmissibility of predictor based on test of significance (regression model). Generalizations.

180. Cohen, A. (1966): All admissible linear estimates of the mean vector. Ann. Math. Statist., 37, 458-463.

The vector Gy is admissible for μ (mean vector) if and only if G is symmetric and all characteristic roots of G lie in [0,1], with at most two roots being equal to 1. If G is asymmetric, then there exists G^* symmetric such that G^*y is better than Gy. If Gy is admissible, then Gy must be generalized Bayes procedure. The generalized prior must be MVN or the product of MVN and uniform distribution.

181. Cohen, A. (1969): A note on the unbiasedness of the likelihood ratio tests for some normal covariance matrices. Sankhya, A, 31, 209-216.

LR tests for various hypotheses concerning Σ matrices. Unbiasedness and Bayes nature of the test. Asymptotic distribution when Σ is diagonal with variances being equal in sets. Particular cases.

Cohen, A. (1971): See Strawderman, W.E. and Cohen, A., #1095.

182. Cole, J.W.L. and Grizzle, J.E. (1966): Applications of multivariate analysis of variance to repeated measurements experiments. Biometrics, 22, 810-828.

MANOVA: its uses and abuses. Rationale and usefulness. Model. Selection of C and M matrices in Roy-Potthoff model. Biological examples. Contrasts with univariate ANOVA.

183. Constantine, A.G. and James, A.T. (1958): On the general canonical correlation distribution. Ann. Math. Statist., 29, 1146-1166.

Distribution of canonical correlations. Direct derivation of Bartlett's results by integration over the extraneous variables. Random vectors in random planes. Averaging over an orthogonal group. Moments of invariants and conditional moments. Tables.

184. Constantine, A.G. (1963): Some noncentral distribution problems in multivariate analysis. Ann. Math. Statist., 34, 1270-1285.

Part expository and part new. Discussion of zonal polynomials. Laplace transforms, hypergeometric functions of matrix argument, integrals. Gamma functions. $P\{S < \Omega\}$. Noncentral Wishart, roots of determinantal equations. Canonical correlations.

185. Constantine, A.G. (1966): The distribution of Hotelling's generalized T_0^2. Ann. Math. Statist., 37, 215-225.

Generalized Laguerre polynomials: generating functions and bounds. Noncentral distribution of T_0^2 in Laguerre polynomials. Moments of T. Tables of $\alpha_{\kappa,\tau}$.

186. Constantine, A.G. and Muirhead, R.J. (1972): Partial differential equations for hypergeometric functions of two argument matrices. J. Multivariate Anal., 2, 332-338.

Derivation of partial differential equations for hypergeometric function of two matrix arguments. No applications are given.

187. Consul, P.C. (1964): Distribution of the determinant of the sum of products matrix in the noncentral linear case. Math. Nachr., 28, 169-179.

Determinant of noncentral Wishart in linear case. Mellin transform methods. Moment function. Integral representation of the density function for general p. Special case of p = 2 (1) 7.

188. Consul, P.C. (1964): On the limiting distribution of some statistics proposed for tests of hypotheses. Sankhya, A, 26, 279-286.

Wilks-Lawley U and Bagai Y statistics. Limiting distributions by Mellin transform methods for p = 2 (1) 8.

189. Consul, P.C. (1965): The exact distribution of certain likelihood criteria useful in multivatiate analysis. Acad. Roy. Belg. Bull. Cl. Sci., 51, 683-691.

Wilks' LR criterion and its hth moment. Mellin transform and its inversion. Method of factorial series. Distribution function and level of significance. Null case.

190. Consul, P.C. (1965): On a new multivariate distribution and its properties. Acad. Roy. Belg. Bull. Cl. Sci., 51, 810-818.

Measure of divergence between two k-variate multinomials. Distribution by characteristic function method. Moment cumulants and distribution of the means. Generalizations. Case of unequal number of traits. Comparison of two different measures.

191. Consul, P.C. (1966): On some inverse Mellin integral transforms. Acad. Roy. Belg. Bull. Cl. Sci., 52, 547-561.

Inverse Mellin transforms for functions occurring in MVN distribution theory: Product of four Gamma functions, product of gwo Gamma quotients, three Gamma quotients. Hypergeometric functions.

192. Consul. P.C. (1966): On some reduction formulae for hypergeometric functions. Acad. Roy. Belg. Bull. Cl. Sci., 52, 562-576.

Some special forms of the hypergeometric function of scalar argument. Particular cases.

193. Consul, P.C. (1966): On the exact distributions of likelihood ratio criteria for testing linear hypotheses about regression coefficients. Ann. Math. Statist., 37, 1319-1330.

Null distribution of $U(p, m, n)$. *Exact representation in an integral form. Inverse Mellin transform. Distribution functions for special values of* p *and* m.

194. Consul, P.C. (1967): On the exact distributions of likelihood ratio criteria for testing independence of sets of variates under the null hypothesis. Ann. Math. Statist., 38, 1160-1169.

Null distribution of the criterion by Mellin transform methods for special values of q, p_1 *and* p_2. *Expressions in terms of hypergeometric functions and G-functions.*

195. Consul, P.C. (1967): On the exact distributions of the criterion W for testing sphericity in a p-variate normal distribution. Ann. Math. Statist., 38, 1170-1174.

Null distribution of W by Mellin transform techniques for $p = 2, 3, 4, 6$. Distribution function. Incomplete Beta function. Hypergeometric function.

196. Consul, P.C. (1968): On some integrals in operational calculus. Bull. Math. (Romanian), 12, 35-42.

Evaluation of an integral and an inverse Mellin transform occurring in multivariate analysis. Hypergeometric function and product of gamma functions.

197. Consul, P.C. (1968): On the distribution of Votaw's likelihood ratio criterion L for testing the bipolarity of a covariance matrix. Math. Nachr., 36, 1-13.

Test of hypothesis that a variance matrix has the bipolar form. Expansion of product of Gamma functions in series. Exact null distribution of L. Method of residues. Coefficients for special cases. Distribution function.

198. Consul, P.C. (1969): On the exact distributions of Votaw's criteria for testing compound symmetry of a covariance matrix. Ann. Math. Statist., 40, 836-843.

Exact null distribution of criterion L for compound symmetry of Σ. Special cases are considered: $p = q = 2$; $p = q = 3$; $p = 3$, $q = 2$; $p = 5$, $q = 2$ or 3. Mellin Transform methods. Distribution function

199. Consul, P.C. (1969): The exact distributions of likelihood ratio criteria for different hypotheses. Multivariate Anal.-II, [Proc. 2nd Internat. Symp., Krishnaiah, ed.], 171-181.

A number of different LR criteria in the null case using Meijer's G function. Mellin transforms: Tests of independence of two or more variates; tests for regression, sphericity and symmetry.

200. Cooper, P.W. (1963): Statistical classification with quadratic forms. Biometrika, 50, 439-448.

Multivariate classification techniques. Optimality of quadratic functions for nonnormal multivariate distributions. Σ's are assumed known. Pearson's Types II and VII.

Cornfield, J. (1963): See Geisser, S. and Cornfield, J., #337.

201. Cornfield, J. (1967): Discriminant functions. Rev. Inst. Internat. Statist., 35, 142-153.

A review article on discriminant functions. Fisher's linear discriminant function, two or more groups. Effects of number of variables. Alternative procedure of classification based on James-Stein estimator.

Cornfield, J. (1967): See Truett, J., Cornfield, J. and Kannel, W., #1140.

202. Cornish, E.A. (1960): The simultaneous fiducial distribution of the parameters of a normal bivariate distribution with equal variances. CSIRO Div. Math. Statist., Tech. Paper #6, 1-8.

Development of joint fiducial distribution of μ_1, μ_2, σ^2 and ρ in BVN with equal variances. Simultaneous inferences based on this distribution.

203. Cornish, E.A. (1961): The simultaneous fiducial distribution of the location parameters in a multivariate normal distribution. CSIRO Div. Math. Statist., Tech. Paper #8, 1-12.

Use of the simultaneous fiducial distribution of location parameters of MVN. Comparison of fiducial and confidence regions using T^2.

204. Cornish, E.A. (1962): The multivariate t-distribution associated with the general multivariate normal distribution. CSIRO Div. Math. Statist., Tech. Paper #13, 1-18.

Joint distribution of $t_i = (x_i - \mu_i)/s_i^{\frac{1}{2}}$ when x is MVN. Derivation by direct integration method. Moments and distributions of linear functions. Distribution of the quadratic forms. Tests of hypotheses. Comparisons with T^2.

Cornish, E.A. (1963): See Bennett, G.W. and Cornish, E.A., #96.

205. Cornish, E.A. (1966): A multiple Behrens-Fisher distribution. Multivariate Anal.-I, [Proc. 1st Internat. Symp., Krishnaiah, ed.], 203-207.

Tests of hypothesis on mean when variance matrices are unequal and unknown. Generalization of Fisher's test for Behrens-Fisher problem. MV t-distribution. Generalization to linear combination of elements of mean vector. Application.

207. Cox, D.R. and Brandwood, L. (1959): On a discriminatory problem connected with the works of Plato. J. Roy. Statist. Soc., B, 21, 195-200.

Problem of discrimination when observations are qualitative and based on frequencies. Classificatory criterion is $\Sigma n_i \ell n(\theta_{1i}/\theta_{0i})$*, weighted total score. Estimation of the criterion. Example: works of Plato.*

208. Cramer, E.M. (1967): Equivalence of two methods of computing discriminant function coefficients. Biometrics, 23, 153.

A simple method of proving Healy's result (#456).

Craswell, K.J. (1964): See Owen, D.B., Craswell, K.J. and Hanson, D.L., #861.

209. Creasy, M.A. (1957): Analysis of variance as an alternative to factor analysis. J. Roy. Statist. Soc. B, 19, 318-325.

Analogy between factor analysis and ANOVA. Approximate tests of significance using F-distribution. Testing and interpretation of interaction. Applications to psychological tests.

210. Curnow, R.N. and Dunnett, C.W. (1962): The numerical evaluation of certain multivariate normal integrals. Ann. Math. Statist., 33, 571-579.

The problem of finding the distribution function of MVN. Representation in two integral forms. Singly or doubly infinite integrals: $\rho_{ij} = \alpha_i \alpha_j$ *or* γ_i/γ_j*. Usefulness of this form of correlation.*

211. Danford, M.B., Hughes, H.M. and McNee, R.C. (1960): On the analysis of repeated-measurements experiments. Biometrics, 16, 547-565.

Analysis of experiments involving repeated measurements on same individual over time. Deviations from assumptions necessary for univariate analysis of variance. Multivariate techniques for testing time and time x treatment interaction effects. Likelihood ratio criterion. MANOVA. Study of serial correlation. Comparison of univariate and multivariate techniques.

212. Daniels, H.E. and Kendall, M.G. (1958): Short proof of Miss Harley's theorem on the correlation coefficient. Biometrika, 45, 571-572.

Short proof of $E(\sin^{-1} r) = \sin^{-1} \rho$*. Sufficiency of* r *for* ρ*.*

213. Darroch, J.N. (1965): An optimal property of principal components. Ann. Math. Statist., 36, 1579-1582.

First k principal components characterized by an optimal property within the class of all random variables. With any matrix A(pxk) and any random vector f(kx1) the quantity $T_1 = trE[(x-Af)(x-Af)']$ *is minimum w.r.t. A and f iff* $Af = \sum_{i=1}^{k} t_i u_i$ *and the minimum value of* $T_1 = \sum_{i=k+1}^{p} \lambda_i$.

214. Darroch, J.N. and Ratcliff, D. (1971): A characterization of the Dirichlet distribution. J. Amer. Statist. Assoc., 66, 641-643.

Characterization: Let $x_1, x_2, \ldots, x_k$ *be such that* $\Sigma x_j < 1$. *Shown that if* $x_i/(1 - \sum_{j \neq i} x_j)$ *is independent of* $\{x_j : j \neq i\}$ *for every* $i = 1, 2, \ldots, k$, *then f(x) has a Dirichlet distribution. Concepts of neutrality and null correlation.*

215. Das Gupta, P. (1968): An approximation to the distribution of sample correlation coefficient, when the population is non-normal. Sankhya, B, 30, 425-428.

Distribution function of $X = \frac{1}{2}(r+1)$, *where r is sample correlation coefficient. Approximation in terms of an incomplete Beta integral and Jacobi polynomials. Constants expressed in terms of the moments of r, which are functions of the bivariate cumulants of parent population. Approximation is exact for BVN with* $\rho = 0$.

216. Das Gupta, S. (1960): Point biserial correlation coefficient and its generalization. Psychometrika, 25, 393-408.

Measuring association between one quantitative and one qualitative character. Properties of sampling distribution (mean and standard error). Tests of hypotheses. Variance-stabilizing transformation. Generalization to point multiserial correlation coefficient.

Das Gupta, S. (1963): See Anderson, T.W. and Das Gupta, S., #36.

Das Gupta, S. (1964): See Anderson, T.W. and Das Gupta, S., #37.

Das Gupta, S. (1964): See Anderson, T.W. and Das Gupta, S., #38.

217. Das Gupta, S., Anderson, T.W. and Mudholkar, G.S. (1964): Monotonicity of the power functions of some tests of the multivariate linear hypothesis. Ann. Math. Statist., 35, 200-205.

Invariant procedures for testing multivariate linear hypotheses depend on characteristic roots of a random matrix $S_h S_e^{-1}$. Sufficient conditions for power function to be a monotonically increasing function of parameters. Test procedures: (i) likelihood-ratio test, (ii) Lawley-Hotelling trace test, (iii) Roy's maximum root test.

218. Das Gupta, S. (1965): Optimum classification rules for classification into two multivariate normal populations. Ann. Math. Statist., 36, 1174-1184. [Correction: 41 (1970), 326].

Known and common covariance matrix: ML rule is an unbiased admissible minimax rule. Unknown and common covariance matrix: ML rule is unbiased and an admissible minimax rule in an invariant class. Loss function is a function of Mahalanobis distance.

219. Das Gupta, S. (1968): Some aspects of discrimination function coefficients. Sankhya, A, 30, 387-400.

Means and covariances of sample discrimination function coefficients. Asymptotic distribution of coefficients. Tests of hypotheses. Monotonicity of power function. Step-down procedures.

220. Das Gupta, S. (1969): Properties of power functions of some tests concerning dispersion matrices of multivariate normal distributions. Ann. Math. Statist., 40, 697-701.

Likelihood ratio tests for (i) $\Sigma = \sigma^2 I$, (ii) $\Sigma = \Sigma_0$ and $\mu = \mu_0$, (iii) $\Sigma = \Sigma_0$. LR test unbiased for (i) and (ii). Modified LR test for (iii) unbiased and power function has monotonicity property. LR test of $\Sigma_1 = \Sigma_2$ is biased when $n_1 \neq n_2$.

221. Das Gupta, S. (1970): A note on some inequalities for multivariate normal distribution. Calcutta Statist. Assoc. Bull., 18, 179-180.

X is MVN with expectation 0 and X_1, X_2 are subvectors. D_1, D_2 are convex sets symmetric about origin.

$$Pr\{X_1 \in D_1, X_2 \in D_2\} \geq Pr\{X_1 \in D_1\} Pr\{X_2 \in D_2\}.$$

221a. Das Gupta, S. (1971): Stepdown multiple decision rules. Essays Prob. Statist., (Roy Volume, Bose et al, eds.), 229-250.

A discussion of stepdown procedures. Reformulation of Anderson's original concepts and theorems. Hotelling's T^2, multiple correlation. Weak unbiasedness.

222. Das Gupta, S. (1971): Non-singularity of the sample covariance matrix. Sankhya, A, 33, 475-478.

Necessary and sufficient conditions in order that $Pr[X(I - J/n)X'$ is singular$] = 0$. Extension to case XMX' where M is any positive semidefinite matrix.

223. Das Gupta, S. (1971): Decomposition of non-null Wilks' U and multivariate Beta. Tech. Rep. #164, School Statist., U. Minnesota.

Bartlett decomposition of Wilks' U in the non-null case. Conditional distribution of sum of products matrix. Application to multivariate Beta (noncentral). Decomposition and distributional properties.

224. Das Gupta, S. (1972): Probability inequalities and errors in classificaton. Tech. Rep. #190, School Statist., U. Minnesota.

Probability inequalities for MVN. Monotonicity property of probabilities of correct classification in two population problems. Estimates of probability of correct classification using minimum distance rules.

225. Das Gupta, S. and Perlman, M.D. (1972): On the power of the noncentral F- and Wilks' U- tests choosing variates for increasing the power of Hotelling's T^2-test. Tech. Rep. #172, School Statist., U. Minnesota.

Discriminant analysis. Power of the test. Noncentrality parameter and its estimation. Rao's covariance adjustment procedure for tests on means. Choice of variates to increase power. Power properties of Wilks' U-test for MANOVA.

David, F.N. (1962): See Barton, D.E. and David, F.N., #80.

226. Davis, A.W. (1967): A counter example relating to certain multivariate generalizations of t and F. Ann. Math. Statist., 38, 613-615.

Multivariate analogues of t and F are dependent on the population covariance matrix. Counter example for assertion (Bennett and Cornish) that distributions are independent of population covariance matrix. Example for $p = 2$, $n = 4$.

227. Davis, A.W. (1968): A system of linear differentgail equations for the distribution of Hotelling's generalized T_0^2. Ann. Math. Statist., 39, 815-832.

Linear differential equations for the null distribution of T_0^2. Laplace transform of $f(T)$. Matrix form of the differential equations. Solution near $T = 0$. Ito's asymptotic expansion for large n_2. Determination of coefficients. Solution near $T = \infty$. Large n_1. Moments. Recurrence relations by use of differential equations.

Davis, A.W. (1968): See Hill, G.W. and Davis, A.W., #463.

228. Davis, A.W. (1970): Further applications of a differential equation for Hotelling's generalized T_0^2. Ann. Inst. Statist. Math., 22, 77-87.

Use of differential equations for computing percentage points. Tables of upper 5 and 1 percentage points. Pillai's trace criterion. Extension of Ito-Siotani expansions. Box's technique. Coefficients up to $O(n_2^{-4})$. Upper percentage points of T_0^2 by this method to $O(n_2^{-3})$. Pillai's criterion. Discussion and comparison with approximations using Pearson's curves.

229. Davis, A.W. (1970): On the null distribution of the sum of the roots of a multivariate beta distribution. Ann. Math. Statist., 41, 1557-1562.

Systems of differential equations satisfied by Pillai's V and relationship with those for T_0^2. Solution to the equations and behaviour in (0, 1). Moments. Ito type expansions. Percentage points to $O(n_2^{-2})$ using this expansion.

230. Davis, A.W. (1970): Exact distributions of Hotelling's generalized T_0^2. Biometrika, 57, 187-191. [Correction: 59 (1972), 498].

Tabulation of the accurate percentiles of T_0^2 for $m = 3$, 4 by using differential equations. Approximations of Ito and Pillai et al. Graphs and tables.

231. Davis, A.W. (1971): Percentile approximations for a class of likelihood ratio criteria. Biometrika, 58, 349-356.

Polynomial expansions for percentile points in the general LR set up using Box's and Fisher-Cornish type expansions. Tables of polynomials. Application of this method to: (i) Wilks' test of independence, (ii) generalized test of homoscedasticity, (iii) Mauchly tests, (iv) Lawley-Hotelling and Pillai's traces, (v) a given mean and covariance matrix. Tables for tests concerning sphericity and H_0: $\Sigma = \Sigma_0$.

232. Davis, A.W. and Field, J.B.F. (1971): Tables of some multivariate test criteria. CSIRO Div. Math. Statist., Tech. Paper #32, 1-21.

Tabulation of upper 5 and 1 percent points using Box-Cornish-Fisher type of expansions. Three cases presented are: (i) Wilks' generalized test of independence, (ii) test of homoscedasicity, (iii) sphericity test.

233. Davis, A.W. (1972): On the marginal distributions of the latent roots of the multivariate Beta matrix. Ann. Math. Statist., 43, 1664-1670.

Differential equations for the marginal distributions derived from those for Hotelling's T_0^2. Solutions. Pillai's approximation.

234. Davis, A.W. (1972): On the distributions of the latent roots and traces of certain random matrices. J. Multivariate Anal., 2, 189-200.

Recursive approach. Laplace transforms. Distributions involving $S_1S_2^{-1}$, $S_1(S_1+S_2)^{-1}$. Marginal distributions of roots of B, largest root of Wishart. Laplace transform of density of V. Differential equations. Null case.

235. Davis, A.W. (1972): On the ratios of the individual latent roots to the trace of a Wishart matrix. J. Multivariate Anal., 2, 440-443.

Relation between the p.d.f. of the ith root of central Wishart and that of the ratio of the root to the trace. Laplace transforms. Null cases.

236. Day, N.E. and Kerridge, D.F. (1967): A general maximum likelihood discriminant. Biometrics, 23, 313-323.

Methods of discriminating when populations are not necessarily MVN. Introduction of disturbance factor $\phi(x)$ which may or may not be known. ML estimation. Reduction to multivariate logistic analyses. Updating the discriminant function.

237. Day, N.E. (1969): Estimating the components of a mixture of normal distributions. Biometrika, 56, 463-474.

Estimation of the components of a mixture of normals (univariate or multivariate). $\Sigma_1 = \Sigma_2 = \Sigma$ (unknown). Estimation procedures: ML, MM, minimum χ^2 and Bayes. Sampling properties of ML estimators investigated by simulation. Application to cluster analysis.

238. Day, N.E. (1969): Linear and quadratic discrimination in pattern recognition. IEEE Trans. Information Theory, 15, 419-421.

Optimum discrimination. Linear discriminant function is optimal for linear exponential family; quadratic discriminant function is optimal in quadratic exponential family.

239. Degroot, M.H. and Li, C.C. (1966): Correlations between similar sets of measurements. Biometrics, 22, 781-790.

Canonical correlation analysis in a restricted sense. Determine coefficient vector a such that a'X and a'Y have maximum correlation when X and Y are identical MVN with zero means, variance matrix Σ *and covariance matrix* Σ_{XY}. *Eigenvectors of the matrix* $\frac{1}{2}\Sigma^{-1}(\Sigma_{XY}+\Sigma'_{XY})$. *ML estimation.*

240. Dempster, A.P. (1958): A high dimensional two sample significance test. Ann. Math. Statist., 29, 995-1010.

Test of $\mu_1 = \mu_2$ *in p-dimensional MVN case when* $n_1+n_2<p+2$. *Hotelling's procedure not applicable. Geometric argument. "Non-exact" significance test based on a test statistic F which is ratio of two mean square distances. Methods for attaching a significance level to F. Approximations for distribution of F.*

241. Dempster, A.P. (1960): A significance test for the separation of two highly multivariate small samples. Biometrics, 16, 41-50.

Application of methods developed in #240. Advantages and disadvantages of techniques. Attaching level of significance by randomization test. Comparison with Chung and Fraser results.

242. Dempster, A.P. (1963): Multivariate theory for general stepwise methods. Ann. Math. Statist., 34, 873-883.

Mathematical basis for stepwise procedures. Independence of tests. Roy's and principal variables techniques. Spherical distributions. Uniqueness. Wishart distribution. Independence. Rao's stepwise LR procedures. Uses and pitfalls of principal variable method.

243. Dempster, A.P. (1963): On a paradox concerning inference about a covariance matrix. Ann. Math. Statist., 34, 1414-1418.

Bayesian or fudicial inference. Posterior distributions. T is Wishart(Σ,n), $P(a) = a'Ta/a'\Sigma a$ *and* $Q(b) = b'\Sigma^{-1}b/b'T^{-1}b$. *Shows that for any linearly invariant posterior distribution of* Σ *given T, cumulative distribution function of P(a) must be uniformly greater than or equal to cumulative distribution function of Q(b). Paradoxical nature of lemma discussed.*

244. Dempster, A.P. (1963): Stepwise multivariate analysis of variance based on principal variables. Biometrics, 19, 478-490.

MANOVA. Stepwise procedure using principal variables. Transformation of data. Uses in case of large number of variables. Tests. Null distributions. Combinations of significance levels. Examples.

245. Dempster, A.P. (1964): Tests for the equality of two covariance matrices in relation to a best linear discriminator analysis. Ann. Math. Statist., 35, 190-199.

Concentration ellipsoid and its shadow property. Configuration of ellipsoids for two sample case under various assumptions on Σ_1, Σ_2. *Relationship between* H_0: $\Sigma_1 = \Sigma_2$ *vs.* H_A: $\Sigma_1 \neq \Sigma_2$ *and linear discriminant function. Tests of hypothesis* H_0: $\Sigma_1 = \Sigma_2$ *for various configurations of ellipsoids. Distribution theory under the null.*

246. Dempster, A.P. (1966): Estimation in multivariate analysis. Multivariate Anal.-I, [Proc. 1st Internat. Symp., Krishnaiah, ed.], 315-333.

Modification of the jack-knife technique to multivariate analysis. Canonical correlation analysis. Deletion of a single degree of freedom. Correction of bias and estimation of sampling variance matrix. Bias correction for future samples.

247. Dempster, A.P. (1971): An overview of multivariate data analysis. J. Multivariate Anal., 1, 316-346.

Multivariate data analysis: its basis and models. Interpretations. Bibliographic.

248. Dempster, A.P. (1972): Covariance selection. Biometrics, 28, 157-175.

Methods for reducing the number of elements to be estimated in a covariance matrix. Exponential family. MLE and uniqueness. Iterative procedures. Examples. Computational theory.

249. Derflinger, G. (1969): Efficient methods for obtaining the Minres and maximum likelihood solutions in factor analysis. Metrika, 14, 214-231.

Analogy between Minres and MLE solution in factor analysis. Rapidly converging iteration algorithms for both methods. Computer programs in FORTRAN IV.

Desu, M.M. (1968): See Geisser, S. and Desu, M.M.; #344.

250. DeWaal, D.J. (1968): An asymptotic distribution for the determinant of a noncentral B statistic in multivariate analysis. South African Statist. J., 2, 77-84.

$L = (A_1+A_2)^{-\frac{1}{2}}A_1(A_1+A_2)^{-\frac{1}{2}}$ where A_1 is central and A_2 is noncentral Wishart. Integral representation of the distribution. Moments of $|L|$ and $|I-L|$. Asymptotic distribution using Box's technique for central and noncentral cases. Noncentrality parameter of rank one (linear case). Distribution of $|I-L|$.

251. DeWaal, D.J. (1969): On the noncentral distribution of the largest canonical correlation coefficient. South African Statist. J., 3, 91-93.

Exact distribution of largest canonical correlation coefficient in the noncentral case. Largest root and its distribution function in a computational form. Case of noncentrality of rank one.

252. DeWaal, D.J. (1969): The noncentral multivariate Beta type 2 distribution. South African Statist. J., 3, 101-108.

$V = B^{-\frac{1}{2}}AB^{-\frac{1}{2}}$ where A is central and B is noncentral Wishart. General p.d.f. involves ${}_0F_1$. Moments of $|V|$ as ${}_1F_1$. If $\Sigma_1 = \Sigma_2 = I$, then V has multivariate Beta type II. If rank of noncentrality is r, then the factorization for this case is established. Distribution of tr V as a double integral. Special cases.

253. DeWaal, D.J. (1970): Distributions connected with a multivariate Beta statistic. Ann. Math. Statist., 41, 1091-1095. [Correction: 42 (1971), 2165-2166].

Integrals involving functions of matrix argument. Multivariate Beta type I integral. Joint distributions of Beta type I matrices. Asymptotic distributions of product of determinants. Box's technique.

254. DeWaal, D.J. (1972): On the expected values of the elementary symmetric functions of a noncentral Wishart matrix. Ann. Math. Statist., 43, 344-347.

Noncentral Wishart. Conjections concerning expected value of the jth elementary symmetric function of roots of a noncentral Wishart for $\Sigma = \sigma^2 I$, and of the pth e.s.f. for any Σ. Expected value of noncentral Wishart matrix and of first e.s.f. Integrals involving two zonal polynomials.

255. DeWaal, D.J. (1972): An asymptotic distribution of noncentral multivariate Dirichlet variates. South African Statist. J., 6, 31-40.

Moments and asymptotic distribution of product of determinants. Complex analogues.

256. Dickey, J.M. (1967): Matricvariate generalizations of the multivariate t distribution and the inverted multivariate t distribution. Ann. Math. Statist., 38, 511-518.

Distribution of T matrix and correlation coefficients. Estimated regression matrix. Marginal, conditional distributions. Submatrices of T. Bayesian applications. Proportional variance matrix case in k population problem. Multivariate Beta.

257. Dickey, J.M. (1968): Three multidimensional-integral identities with Bayesian applications. Ann. Math. Statist., 39, 1615-1627.

Three identities concerning MV distributions: two of these are about MV t distribution and one on Dirichlet distribution. Moments of products of MV t-variates. MV Behrens-Fisher problem and distribution. Inferences concerning MVN location parameters. Likelihood function and inferences under Student t distribution of errors.

258. Dickey, J.M. (1971): The weighted likelihood ratio, linear hypotheses on normal location parameters. Ann. Math. Statist., 42, 204-223.

Conjugate prior distributions and Jeffrey's theory. Sharp hypothesis. MVN extensions. Linear models. W-orthogonality. MV Behrens-Fisher problem and its solution.

259. DiDonato, A.R. and Jarnagin, M.P. (1961): Integration of the general bivariate Gaussian distribution over an offset circle. Math. Comput., 15, 375-382.

General BVN. Evaluation of the integral. Computation of error function and its derivation. Tables.

DiDonato, A.R. (1961): See Weingarten, H. and DiDonato, A.R., #1163.

260. DiDonota, A.R. and Jarnagin, M.P. (1962): A method for computing the circular coverage function. Math. Comput., 16, 347-355.

Circular normal distributions. Numerical evaluation of offset circle. Recursive relations. Errors of approximations. Tables.

261. Doktorov, B.Z. (1969): Some estimators for the multidimensional normal distribution. Theory Prob. Appl., 14, 526-528.

Estimation in MVN under various assumptions: (i) σ_1 of the first component, all other parameters known except μ. Asymptotic variance of $\hat{\sigma}_1$, MLE. (ii) $p = 2$ case discussed in detail. Table of estimation of σ_1 under various assumptions along with variance of estimator $\hat{\sigma}_1$. (iii) Estimation of ρ. Applications.

Dotson, C.O. (1969): See Pillai, K.C.S. and Dotson, C.O., #891.

Dubman, M.R. (1969): See Goodman, N.R. and Dubman, M.R., #391.

262. Dubois, N.S.D'Andrea, Jr. (1969): A solution to the problem of linking multivariate documents. J. Amer. Statist. Assoc., 64, 163-174.

Document linkage problem. Classifying pairs of documents into one of two populations. Items take on three values: correct, incorrect, missing. Classification procedures. ML estimation. Probability of misclassification.

263. Dudewicz, E.J. (1969): An approximation to the sample size in selection problems. Ann. Math. Statist., 40, 492-497.

Approximate solution for δ such that $Pr\{X > -\delta\} = \alpha$ in standardized MVN. Application to problem of obtaining sample size in a selection procedure of Bechhofer. Uses in problem of selection from a MVN population.

264. Dunn, O.J. (1958): Estimation of the means of dependent variables. Ann. Math. Statist., 29, 1095-1111.

Simultaneous confidence intervals for means of MVN variables. Exact confidence coefficient type: fixed length for known variance, varying length for unknown but equal variances. Bounded confidence coefficient type. Information about covariances is not needed.

265. Dunn, O.J. (1959): Confidence intervals for the means of dependent normally distributed variables. J. Amer. Statist. Assoc., 54, 613-621.

Illustrates techniques developed in #264. Confidence intervals (i) variances known (correlation unspecified), (ii) variances unknown but equal, (iii) variances unknown and unequal. Numerical examples.

266. Dunn, O.J. (1961): Multiple comparisons among means. J. Amer. Statist. Assoc., 56, 52-64.

Simultaneous confidence intervals for m linear combinations (as opposed to contrasts) among k means whose variance matrix is known except for a multiplicative constant σ^2. Methods used: t distribution, F distribution, Tukey's method.

267. Dunn, O.J. (1965): A property of the multivariate t distribution. Ann. Math. Statist., 36, 712-714.

Student's t distribution evaluated from -u to u is an increasing function of its d.f. Monotonicity property not valid for multivariate t distribution.

268. Dunn, O.J. and Massey, F.J., Jr. (1965): Estimation of multiple contrasts using t distributions. J. Amer. Statist. Assoc., 60, 573-583.

Simultaneous confidence intervals using multivariate t-distribution. Review of available tables. Approximate techniques. Comparison of exact and approximate results for equicorrelated case.

269. Dunn, O.J. and Varady, P.D. (1966): Probabilities of correct classification in discriminant analysis. Biometrics, 22, 908-924.

Monte Carlo techniques. Comparison of actual probability of correct classification using a linear discriminant function and the estimate of this probability obtained by estimating the Mahalanobis distance. Graphs giving approximate conservative intervals for the probabilities of correct classification.

Dunn, O.J. (1966): See Weiner, J.M. and Dunn, O.J., #1162.

Dunn, O.J. (1967): See Holloway, L.N. and Dunn, O.J., #470.

270. Dunn, O.J. and Clark, V. (1969): Correlation coefficients measured on the same individuals. J. Amer. Statist. Assoc., 64, 366-377.

The quadivariate normal. Tests for H_0: $\rho_{12} = \rho_{13}$ and H_0: $\rho_{12} = \rho_{34}$ based on transformed variable Z: Five assymptotic tests for these hypotheses are discussed and compared. Graphical comparison. Discussion of power.

271. Dunn, O.J. (1971): Some expected values for probabilities of correct classification in discriminant analysis. Technometrics, 13, 345-353.

Monte Carlo study of (i) unconditional probability of correct classification, (ii) expected value of its estimate based on the calculated Mahalanobis distances. Lachenbuch's approximation. Graphical and tabular display of results.

272. Dunn, O.J. and Clark, V. (1971): Comparison of tests of the equality of dependent correlation coefficients. J. Amer. Statist. Assoc., 66, 904-908.

Continuation of #270. Monte Carlo (small sample) study of four procedures for testing H_0: $\rho_{12} = \rho_{13}$ and four others for H_0: $\rho_{12} = \rho_{34}$. These are all studied from α-level point of view. Graphs. Discussion.

Dunn, O.J. (1972): See Chan, L.S. and Dunn, O.J., #146.

273. Dunnett, C.W. (1960): On selecting the largest of k normal population means. J. Roy. Statist. Soc., B, 22, 1-40.

k variate normal. Selection of variate with largest mean. Σ known and of the form $\sigma^2[(1-\rho)I+\rho J]$. Prior of μ is MVN with means known to a constant and variance matrix known. Independence assumed in development of results. Probability of correct selection. Conditional probability of a correct selection. Probability that the selected mean is within a specified amount of the largest mean. Linear less. Sample sizes. Expected value of the largest of k normal variates equally, positively correlated, having same variance but different means. Discussion by various statisticians.

Dunnett, C.W. (1962): See Curnow, R.N. and Dunnett, C.W., #210.

274. Dunsmore, I.R. (1966): A Bayesian approach to classification. J. Roy. Statist. Soc., B, 28, 568-577.

Prediction. Bayesian classification: σ^2 known and σ^2 unknown. Conjugate priors: normal and Dirichlet. Examples to real life situations. Discussion of classical techniques for univariate case.

275. Dwyer, P.S. (1967): Some applications of matrix derivatives in multivariate analysis. J. Amer. Statist. Assoc., 62, 607-625. [Correction: 62 (1967), 1518].

Differentiation of matrix functions. Formulae summarized in tables. Applications: ML estimates for complex likelihood functions, optimizing matrix functions when there are matrices of side conditions, evaluating Jacobians.

Dwyer, P.S. (1969): See Tracy, D.S. and Dwyer, P.S., #1131.

276. Dykstra, R.L. (1970): Establishing the positive definiteness of the sample covariance matrix. Ann. Math. Statist., 41, 2153-2154.

Sample covariance matrix is positive definite with probability one if and only if $N > p$ where p is dimensionality of MVN and N is the sample size.

277. Eagleson, G.K. (1964): Polynomial expansions of bivariate distributions. Ann. Math. Statist., 35, 1208-1215.

Class of bivariate distributions whose canonical variables are the orthogonal polynomials of the marginals. BVN, Gamma, Poisson, negative binomial and hypergeometric distributions. Regression properties. Correlation as a measure of dependence. Hermite-Chebychev polynomials.

278. Eagleson, G.K. (1968): A note on linear regression in trivariate distributions. J. Amer. Statist. Assoc., 63, 1042-1044.

Additive family and additive random elements in common. Linearity of regression of one marginal on other marginals. Sufficient conditions. TVN and some discrete distributions.

279. East, D.A. and Oschinsky, L. (1958): A comparison of serological and somatometrical methods used in differentiating between certain East African racial groups, with special reference to D^2 analysis. Sankhya, 20, 31-68.

An application of D^2 (generalized distance) analysis to anthropometric data from East Africa. Selection of variates to discriminate between closely related groups. Tests of normality. Complete data appended.

280. Eaton, M.L. (1966): Characterization of distributions by the identical distribution of linear forms. J. Appl. Prob., 3, 481-494.

A general discussion of characterizations of MV distributions. Infinite divisibility. Multivariate normality based on the distribution of $\Sigma X_i B_i$. Other stable laws.

281. Eaton, M.L. (1967): The generalized variance: Testing and ranking problem. Ann. Math. Statist., 38, 941-943.

Monotone likelihood ratio character of the density function of the generalized variance. UMPI tests of the hypothesis H_0: $|\Sigma| \leq c_1$ *against* H_1: $|\Sigma| > c_2$ *where* $0 < c_1 \leq c_2$. *Maximum and most stringent test. Ranking of several MVN distributions on the basis of generalized variance. Properties of decision rule based on the generalized variance.*

282. Eaton, M.L. (1969): Some remarks on Scheffe's solution to the Behrens-Fisher problem. J. Amer. Statist. Assoc., 64, 1318-1322.

Generalization of Bennett-Anderson solution to MV Behrens-Fisher problem: k-samples each consisting of different numbers of elements by the use of T^2*-test. Null and non-null distributions of test statistic.*

283. Eaton, M.L. and Efron, B. (1970): Hotelling's T^2 test under symmetry conditions. J. Amer. Statist. Assoc., 65, 702-711.

Hypothesis of orthant symmetry: generalization of normality and of equal variances. Behaviour of T^2 *under orthant symmetry (rather than the usual normality conditions). Bounds on probability of type I error. Use of* T^2 *to test orthant symmetry. Nonparametric version of* T^2*: sign test. Null distribution of* T^2 *from geometric considerations.*

284. Eaton, M.L. and Olshen, R.A. (1972): Random quotients and the Behrens-Fisher problem. Ann. Math. Statist., 43, 1852-1860.

Generalizations of Lawton's results. Exchangeable random matrices and majorization. Probability inequalities. Extension of Hsu's results. MV Behrens-Fisher problem.

285. Eaton, M.L. and Perlman, M.D. (1972): A monotonicity property of the power functions of some invariant tests for MANOVA. Tech. Rep. #193, School Statist., U. Minnesota.

The main result is that the power function $\rho_k(\delta)$ *is a Schur-convex function of* δ *if k is a convex set in (X,S). Here X and S refer to the usual MANOVA set up. This provides a result regarding the behaviour of the power function along contours linear in* δ. *Monotonicity of the power of invariant tests. Confirmation of Pillai-Jayachandran numerical results and of Fujikoshi concerning power behaviour of MANOVA tests.*

286. Edwards, A.W.F. and Cavalli-Sforza, L.L. (1965): A method for cluster analysis. Biometrics, 21, 362-375.

Investigation of relationships of points in multidimensional space. ANOVA techniques. Applications to classification. Sequential building up of trees. Comparison and relationship to discriminant analysis. Examples.

287. Edwards, A.W.F. (1971): Distances between populations on the basis of gene frequencies. Biometrics, 27, 873-881.

A rebuttal to Balakrishnan and Sanghvi paper (#67). Angular transformations. Stereographic projections. Co-ordinates of points and distances between points. Irrelevance of correlations to distance.

Efron, B. (1970): See Eaton, M.L. and Efron, B., #283.

288. Efron, B. and Morris, C. (1972): Empirical Bayes on vector observations: An extension of Stein's method. Biometrika, 59, 335-347.

Construction of an estimator which dominates MLE in MVN case in the sense of risk for squared error loss (normalized). Normal prior for the means. Relative savings loss. Estimation of the variance matrix. Use of James-Stein rule for each co-ordinate. Unknown prior mean. Modification of the decision rule. Applications to two-way classification. Mandel's procedure. Improvement introduced by James-Stein procedures.

289. Efron, B. and Morris, C. (1972): Limiting the risk of Bayes and empirical Bayes estimators-Part II: The empirical Bayes case. J. Amer. Statist. Assoc., 67, 130-139.

Compromise estimation between MLE and James-Stein procedure. Limited translation estimators of #288 considered in empirical Bayes set up. Relationship of James-Stein procedure with the empirical Bayes procedure. Ensemble and maximum component risks. Estimation of union functions of the mean. Noncentrality of the prior.

290. Elashoff, J.D., Elashoff, R.M. and Goldman, G.E. (1967): On the choice of variables in classification problem with dichotomous variables. Biometrika, 54, 668-670.

Classification into one of two populations on the basis of dichotomous variables. Extension of Cochran's results for MVN case. Choice of two out of p dichotomous variables. Probability of misclassification. Variables are independent: k best chosen separately not generally the best k chosen k at a time. Variables dependent: positive correlation may increase discrimination and negative correlation may decrease discrimination. Stepwise selection rule for choosing best subset.

Elashoff, R.M. (1966): See Afifi, A.A. and Elashoff, R.M., #3.

Elashoff, R.M. (1967): See Afifi, A.A. and Elashoff, R.M., #4.

Elashoff, R.M. (1967): See Elashoff, J.D., Elashoff, R.M. and Goldman, G.E., #290.

Elashoff, R.M. (1969): See Afifi, A.A. and Elashoff, R.M., #5.

Elashoff, R.M. (1969): See Afifi, A.A. and Elashoff, R.M., #6.

Elashoff, R.M. (1969): See Afifi, A.A. and Elashoff, R.M., #7.

291. Ellison, B.E. (1962): A classification problem in which information about alternative distributions is based on samples. Ann. Math. Statist., 33, 213-223.

Methods for obtaining translation invariant Bayes procedures for a classification problem. Minimum distance and restricted maximum likelihood procedures. Admissibility. Normal priors and 0-1 loss.

292. Ellison, B.E. (1965): Multivariate-normal classification with covariance unknown. Ann. Math. Statist., 36, 1787-1793.

Generalization of #291 when the means are linearly restricted.

293. Elston, R.C. and Grizzle, J.E. (1962): Estimation of time-response curves and their confidence bands. Biometrics, 18, 148-159.

Exemplification of Rao's techniques in the analysis of growth curves. Data analyzed in three ways: (i) MV model, using Rao's method, (ii) complete independence model, (iii) mixed effects model. Discussion of the results.

294. Enis, P. and Geisser, S. (1970): Sample discriminants which minimize posterior squared error loss. South African Statist. J., 4, 85-93.

There are two broad classes of problems considered: (i) $\Sigma_1 \neq \Sigma_2$, quadratic case, (ii) $\Sigma_1 = \Sigma_2$, linear case. Under (i) there are a number of variants: (a) all parameters unknown, (b) Σ_1, Σ_2 known, (c) Σ_1, Σ_2 unknown, μ_1, μ_2 known, (d) Σ_1, Σ_2 unknown but $\mu_1 = \mu_2 = \mu$ known, (e) Σ_1, Σ_2 known but $\mu_1 = \mu_2 = \mu$ unknown. In case (ii) there are two considered: (a) Σ unknown, μ_1, μ_2 known, (b) Σ known, μ_1, μ_2 unknown. The priors in all cases are usual non-informative priors.

295. Enis, P. and Geisser, S. (1971): Estimation of the probability that $Y < X$. J. Amer. Statist. Assoc., 66, 162-168.

$P\{Y < X\}$ from a parametric and Bayesian view-point. Predictive probability as the mean of the posterior distribution. BVN: Independent and dependent cases. Known and unknown parameters. Non-information priors. MVN: Estimation of $P\{\Sigma b_i X_i > 0\}$.

296. Evans, I.G. (1965): Bayesian estimation of parameters of a multivariate normal distribution. J. Roy. Statist. Soc., B, 27, 279-283.

Estimation of μ, Σ and $|\Sigma|$ in MVN using Bayesian procedures. The loss function of a general form. Natural conjugates as priors. Cases of (i) μ known, Σ unknown, (ii) μ unknown, Σ known, (iii) both unknown. Discussion.

Fairchild, M.D. (1972): See Jenden, D.J., Fairchild, M.D., Mickey, M.R., Silverman, R.W. and Yale, C., #513.

Fallis, R.F. (1965): See Wherry, R.J., Naylor, J.C., Wherry, R.J., Jr., and Fallis, R.F., #1165.

Feiveson, A.H. (1966): See Odell, P. and Feiveson, A.H., #843.

297. Feldman, S., Klein, D.F. and Honigfeld, G. (1969): A comparison of successive screening and discriminant function techniques in medical taxonomy. Biometrics, 25, 725-734.

Use of successive screening, for qualitative group assignment, in place of discriminant analysis. Maximization of the positive to false-positive ratio. Psychiatric traits. Virtues of screening procedure and weaknesses of discriminant function. No normality required.

Feldt, L.S. (1970): See Huynh, H. and Feldt, L.S., #483.

Field, J.B.F. (1971): See Davis, A.W. and Field, J.B.F., #232.

Fienberg, S. (1969): See Tiao, G.C. and Fienberg, S., #1126.

298. Fiering, M.B. (1962): On the use of correlation to augment data. J. Amer. Statist. Assoc., 57, 20-32.

Missing data in one variate for BVN and TVN. Estimation of missing observations by regression. Effects of the procedure on $\hat{\mu}$ and $Var(\hat{\mu})$. Optimal values of ρ^2. Cost function studies.

299. Finney, D.J. (1962): Cumulants of truncated multinormal distributions. J. Roy. Statist. Soc., B, 24, 535-536.

Cumulants generating functions for standardized MVN when truncated at arbitrary points on all co-ordinates. Cumulants.

300. Fisher, G. (1962): A discriminant analysis of reporting errors in health interviews. Appl. Statist., 11, 148-163.

Application of discriminant function for prediction in US Naval health data. Hospitalization by various criteria.

301. Fisher, R.A. (1962): The simultaneous distribution of correlation coefficients. Sankhya, A, 24, 1-8.

Simultaneous distribution of r_{ij} expressed in integral form. Geometrical interpretation.

302. Fisk, P.R. (1970): A note on a characterization of the multivariate normal distribution. Ann. Math. Statist., 41, 486-494.

NSC for the joint distribution of a partitioned vector to be MVN. Linearity of regression of one component on the other. Generalizations to several partitions.

303. Fleiss, J.L. (1966): Assessing the accuracy of multivariate observations. J. Amer. Statist. Assoc., 61, 403-412.

Reliability of multivariate observations. Similarity with multivariate linear model. ML estimation of mean vector and variance matrix. Assessing accuracy by degree of agreement among observers: LR test on mean vector. Index of reliability in the p-variate case. Repeated measurements problems.

304. Foote, R.J. (1958): A modified Doolittle approach for multiple and partial correlation and regression. J. Amer. Statist. Assoc., 53, 133-143.

Multiple and partial correlation and regression coefficient evaluation by use of desk calculators. Coding, interchanging, addition or elimination of variables. Standard errors and other computation.

305. Foster, F.G. and Rees, D.H. (1957): Upper percentage points of the generalized Beta distribution. I. Biometrika, 44, 237-244.

Tabulation of the largest root of $|B-\lambda A| = 0$. Extension of tables of Pillai and Nanda. Application to MANOVA, regression. Interpolation procedure. Tables.

306. Foster, F.G. (1957): Upper percentage points of the generalized Beta distribution. II. Biometrika, 44, 441-443.

Tables extended to 80, 85, 90, 95 and 99 percent upper points of $I_x(k; p, q)$ for $k = 3$. Method of computation. Example.

307. Foster, F.G. (1958): Upper percentage points of the generalized Beta distribution. III. Biometrika, 45, 492-493.

Extension of #305 and #306 to $k = 4$.

308. Fox, C. (1957): Some applications of Mellin transforms to the theory of bivariate statistical distributions. Proc. Cambridge Philos. Soc., 53, 620-628.

Two dimensional Mellin transforms. Inverse transform to determine the distribution of $|x|^{r-1}|y|^{s-1}$. Distributions of products and quotients of bivariate random variables. Symmetrical and nonsymmetrical distributions.

Fox, K.A. (1957): See Waugh, F.V. and Fox, K.A., #1158.

309. Fraser, A.R. and Kovats, M. (1966): Stereoscopic models of multivariate statistical data. Biometrics, 22, 358-367.

Plotting of three dimensional data. Co-ordinates. Example.

Fraser, D.A.S. (1958): See Chung, J.H. and Fraser, D.A.S., #170.

310. Fraser, D.A.S. (1968): The conditional Wishart: normal and nonnormal Ann. Math. Statist., 39, 593-605.

Affine multivariate model and the distributions of the model. Error symmetry. Relationship to Wishart distribution. Linear multivariate model. Error symmetry and central Wishart. Noncentral Wishart. Conditional. Structural model.

311. Fraser, D.A.S. and Haq, M.S. (1969): Structural probability and prediction for the multivariate model. J. Roy. Statist. Soc., B, 31, 317-331.

Structural distribution for Σ and μ in MV case. MVN generated by specializing rotationally symmetric error to a normal distribution. Relationship to T^2 and to posteriors of the Bayesians. Problem of prediction of $\bar{x}$ and S and of a set of variables.

312. Freeman, H., Kuzmack, A. and Maurice, R. (1967): Multivariate t and the ranking problem. Biometrika, 52, 305-308.

Ranking of 3, 4 or 5 normal populations. Determination of sample size required. Use of MV t. Minimum values of expected sample size. Computational procedures.

313. Freeman, H. and Kuzmack, A.M. (1972): Tables of multivariate t in six and more dimensions. Biometrika, 59, 217-219.

Tables. Percentage points for 90, 95, and 99 percent. Upper tail. $k = 6, 9, 10, 15, 20, 25$ and 30 and various sample sizes (initial).

314. Freund, J.E. (1961): A bivariate extension of the exponential distribution. J. Amer. Statist. Assoc., 56, 971-977.

Properties of BV exponential. Moments, regressions of Y on X and of X on Y. MLE. Mean and variances of the estimators.

Freund, R.J. (1962): See Appleby, R.H. and Freund, R.J., #46.

315. Friedman, H.P. and Rubin, J. (1967): On some invariant criteria for grouping data. J. Amer. Statist. Assoc., 62, 1159-1178.

Discussion of methods of determining the clustering of data. Total scatter, maximization of $|T|/|W|$, maximization of $tr(W^{-1}B)$. Pairwise distance criteria. Computational techniques. Case of singular matrices and small number of observations. Discriminant analysis. Graphical representations. Application to rela and artificial data. Bibliographic.

316. Fujikoshi, Y. (1968): Asymptotic expansion of the distribution of the generalized variance in the noncentral case. J. Sci. Hiroshima Univ., Ser. A-1, 32, 293-299.

MGF of sample variance matrix S. Noncentral. Asymptotic distribution of $\sqrt{n}\,[f\{\Sigma^{-\frac{1}{2}}S\Sigma^{-\frac{1}{2}}\} - f(I+\theta)]$. Asymptotic distribution of $|S|$ when noncentrality parameter is $O(n)$ and when it is $O(1)$. Characteristic function method.

Fujikoshi, Y. (1969): See Sugiura, N. and Fujikoshi, Y., #1106.

317. Fujikoshi, Y. (1970): Asymptotic expansions of the distributions of test statistics in multivariate analysis. J. Sci. Hiroshima Univ., Ser. A-1, 34, 73-144.

Zonal polynomials. Laguerre polynomials and Laplace transforms involving matrix argument. Asymptotic expansions of Pillai's $V^{(p)}$, Hotelling's T_0^2. Tests of independence. LR criteria for H_0: $\mu_i = \mu_.$, $\Sigma_i = \Sigma_.$ against $\mu_i \neq \mu_.$, $\Sigma_i = \Sigma_.$. Generalized variance and trace of noncentral Wishart. Characteristic functions. Extensive discussion.

318. Fujikoshi, Y. (1971): Asymptotic expansions of the non-null distributions of two criteria for the linear hypothesis concerning complex multivariate populations. Ann. Inst. Statist. Math., 23, 477-490.

Complex analogues of #317 regarding Pillai's and Hotelling's criteria for linear hypotheses. Zonal polynomials. Asymptotic expansions. MGF. Differential operators.

319. Fujikoshi, Y. (1972): Asymptotic formulas for the distributions of the determinant and trace of a noncentral Beta matrix. J. Multivariate Anal., 2, 208-218.

Hypergeometric function of matrix argument. Asymptotic expansion of characteristic function of $|B|$ and tr B when B is noncentral beta matrix. Criteria arising in MV linear hypothesis. Assumption $\Sigma = I$ and Ω of rank p.

Fujisawa, H. (1966): See Kudo, A. and Fujisawa, H., #692.

320. Fukutomi, K. and Sugiyama, T. (1967): On the distributions of the extreme characteristic roots of the matrices in multivariate analysis. Rep. Statist. Appl. Res., JUSE, 14, 8-12.

Generalized Beta and F (MV). Central case. Joint density of the roots. Marginals of the extreme roots (largest or smallest). Direct integration method. Hypergeometric function of matrix argument. Singular case ($n_1 < p$). Computation of the upper percentage points.

321. Furukawa, N. (1962): The inference theory in multivariate random effect model, I. Kumamoto J. Sci. Ser. A, 5, 158-170.

Random effects model with observations as well as treatment effects MVN. NSC for covariance matrices to be estimable. Completeness of family of distributions of the sufficient statistics.

322. Gabriel, K.R. (1962): Ante-dependence analysis of an ordered set of variables. Ann. Math. Statist., 33, 201-212.

Ante-dependence and analogy with Markov chains. Partial correlation and characterization of ante-dependence. Test for the partial correlation being zero. LR. Partial independence. MLE. Asymptotic distribution of criterion. Multidimensional generalizations.

323. Gabriel, K.R. (1968): Simultaneous test procedures in multivariate analysis of variance. Biometrika, 55, 489-504.

MANOVA and simultaneous test procedures. Coherence and consonance of STP. Maximum root, Hotelling's T_0^2, Pillai's trace, Wilks' LR criteria. SC bounds. Minimal rejected combinations of subsets and variables. General MV linear hypothesis. Power. Inequalities on roots of some matrices and their functions. MANOVA. STP properties: proofs of the properties.

324. Gabriel, K.R. (1969): Simultaneous test procedures - some theory of multiple comparisons. Ann. Math. Statist., 40, 224-250.

Review and bibliographic. Families of hypotheses and statistics. Ordering and monotonicity. Simultaneous test procedures (STP) and confidence statements (SCS), and their relation with LR. STP and coherence, consonance and monotonicity. Parsimony and resolution. Analogies in the properties of STP and SCS. Methods of multiple comparisons. MANOVA. Examples.

325. Gabriel, K.R. (1969): A comparison of some methods of simultaneous inference in MANOVA. Multivariate Anal.-II, [Proc. 2nd Internat. Symp., Krishnaiah, ed.], 67-86.

Tabular representation of simultaneous inference procedures. STP and SCB in MANOVA. Stepdown procedures. Bounds on deviations from atomic hypotheses of Ω. Distributions of statistics. Coherence and consonance. Resolution. Example of trivariate, six class, one-way MANOVA.

326. Gabriel, K.R. (1970): On the relation between union intersection and likelihood ratio tests. Essays Prob. Statist., (Roy Volume, Bose et al, eds.), 251-256.

LR and UI methods. Conditions for LR tests to induce LRUI critical region that is contained in LR region. Equality of LRUI and LR tests and statistics. MANOVA. Rank of the hypothesis. Hotelling's T^2. Roy's maximum root test. STP, SLR. Power considerations. A criterion of Darroch and Silvey.

327. Gabriel, K.R. (1971): The biplot graphic display of matrices with application to principal component analysis. Biometrika, 58, 453-467.

Factorization of a matrix. Graphical display. Approximate biplot of any matrix by approximating elements of Y by nonsingular decomposition. Principal component biplot. Distance (standardized) between points. Canonical decomposition. Goodness of fit. Extensions to higher dimensions.

328. Gajjar, A.V. (1967): Limiting distributions of certain transformations of multiple correlation coefficient. Metron, 26, 189-193.

Distribution of $\tanh^{-1} R$. *Approximation to distribution of* R^2 *by use of Stirling's approximation. Distribution of* $h(\rho)R^2/(1-R^2)$ *and* $h_1(\rho)R^2/(1-R^2)$ *approximated by noncentral and central F distributions.* $\sin^{-1} R$.

329. Gales, K. (1957): Discriminant functions of socio-economic class. Appl. Statist., 6, 123-132.

Assignment of individuals to one of three socio-economic classes by discriminant analysis. Misclassifications. Levels of agreement between different classificatory techniques. Example with eight attributes.

330. Gardner, M.J. (1972): On using an estimated regression line in a second sample. Biometrika, 59, 263-274.

Prediction of dependent variable using second sample. A measure of excess of the sum of squares about the line compared with the least squares line. Bias. Regressor variables from MVN. Expectation and variance of the bias. Properties of the elements of an inverse Wishart matrix.

Garg, M.L. (1968): See Rao, B.R., Garg, M.L. and Li, C.C., #925.

331. Gatty, R. (1966): Multivariate analysis for marketing research: an evaluation. Appl. Statist., 15, 157-172.

Applications of multivariate analyses to marketing research. Multiple regression, multiple correlation. Factor and principal component analysis. MANOVA and covariance techniques. Discussion. Bibliographic.

Gaudin, M. (1965): See Mehta, M.L. and Gaudin, M., #789.

332. Gehan, E.A. (1962): An application of multivariate regression analysis to the problem of predicting survival in patients with severe lukemia. Bull. Inst. Internat. Statist., 39, 173-179.

Multiple regression using partial correlations to rank the variables in order of importance. Standard errors of predicted value of survival discussed and tabled.

333. Geisser, S. and Greenhouse, S.W. (1958): An extension of Box's results on the use of the F distribution in multivariate analysis. Ann. Math. Statist., 29, 885-891.

MV approach to mixed model in two-way classification with equally correlated errors. Extension of Box's results to g groups. Approximate F-tests. Joint test of groups and treatment x group interaction. Unequal covariance matrices for g groups.

334. Geisser, S. (1959): A method for testing treatment effects in the presence of learning. Biometrics, 15, 389-395.

Analysis of a Latin square by multivariate techniques. Correlated observations. Testing equality of treatment means using T^2. Additive model and equal covariance matrices. Analysis of repeated measurements experiments.

Geisser, S. (1959): See Greenhouse, S.W. and Geisser, S., #401.

335. Geisser, S. and Mantel, N. (1962): Pairwise independence of jointly dependent variables. Ann. Math. Statist., 33, 290-291.

Sample correlations based on p independent non-singular normal variates. Exhibit pairwise independence but not mutual independence. Application to computation of variance of linear combinations of correlation coefficients.

336. Geisser, S. (1963): Multivariate analysis of variance for a special covariance case. J. Amer. Statist. Assoc., 58, 660-669. [Correction: 59 (1964), 1296].

MANOVA when $\Sigma = \sigma^2[(1-\rho)I+\rho J]$. Analogy to mixed models in which errors are serially correlated. Test of $\mu = \mu_0$. Extension to equality of vector means of g populations. Linear function of independent F statistics. Patterned covariance matrices.

337. Geisser, S. and Cornfield, J. (1963): Posterior distributions for multivariate normal parameters. J. Roy. Statist. Soc., B, 25, 368-376.

Bayesian posterior distributions for parameters of MVN. Comparison with fiducial and confidence counterparts. No single value of parameter of prior distribution will lead to: (i) t as distribution for a marginal mean and (ii) T^2 for the joint distribution of all means.

338. Geisser, S. (1964): Posterior odds for multivariate normal classifications. J. Roy. Statist. Soc., B, 26, 69-76.

Bayesian procedures. Posterior probability of an observation belonging to one of k MVN populations. Assumptions on the μ_i, Σ_i: known, unknown, equal, unequal (nine such combinations). The case of the proportions $p_1,\ldots,p_k$ unknown is considered. Examples.

339. Geisser, S. (1964): Estimation in the uniform covariance case. J. Roy. Statist. Soc., B, 26, 477-483.

Basic methods of interval estimation (regions). Consistency with one another. BVN with equal marginal variances. Confidence, Bayesian and fiducial regions. Uniform prior. MVN case with equal variances and covariances. Posterior of r.

340. Geisser, S. (1965): Bayesian estimation in multivariate analysis. Ann. Math. Statist., 36, 150-159.

Posterior distributions of MV statistics with priors reflecting ignorance or relative diffuseness. Bayesian interpretation of standard procedures. Bayesian intervals for μ and for linear combinations of μ when Σ is not known. Ratio of means for BVN. Posterior densities for the inverse of Σ and its functions. Principal components and roots. Several population cases. Joint region for $\mu_1,\ldots,\mu_k$ and the region for the linear function of these means. Behrens-Fisher problem and approximation. General linear hypothesis. Posterior of β and inverse of Σ. Posterior region on elements of β. Equivalence with the confidence region. Predictive regions.

341. Geisser, S. (1965): A Bayes approach for combining correlated estimates. J. Amer. Statist. Assoc., 60, 602-607.

MVN with arbitrary variance matrix and all elements of mean vector equal to μ. Bayesian solutions to interval estimation using (i) diffuse prior, (ii) natural conjugate prior. Student's t distribution. Parallel profile analysis: two sample case.

342. Geisser, S. (1966): Predictive discrimination. Multivariate Anal.-I, [Proc. 1st Internat. Symp., Krishnaiah, ed.], 149-163.

Discrimination by Bayesian procedures with known prior probabilities. Predictive distribution and classification. MVN discrimination for a variety of combinations of μ, Σ being known or unknown. Equality of covariance matrices. Predictive linear discriminators. Joint normal classification. Sequential classification procedures. Comparison and discussion.

343. Geisser, S. (1967): Estimation associated with linear discriminants. Ann. Math. Statist., 38, 807-817.

Anderson-Wald linear discriminant function. Bayesian analysis using diffuse prior. Estimation of the population discriminant, true errors of misclassification, errors incurred by use of sample discriminant on future observations.

344. Geisser, S. and Desu, M.M. (1968): Predictive zero-mean uniform discrimination. Biometrika, 55, 519-524.

Discrimination in zero mean, unequal variance matrix case. Variance matrix of ith population of the form $\sigma_i^2[(1-\rho_i)I+\rho_i J]$*. Cases considered: (i)* ρ_i, σ_i *both known, (ii)* ρ_i *known but* σ_i *are unknown, (iii)* $\sigma_1 = \sigma_2 = \sigma$ *unknown,* $\rho_1 \neq \rho_2$ *both known, (iv) all four unknown. Posterior distributions. Genetic example. Comparison with Bartlett-Please results.*

345. Geisser, S. (1970): Bayesian analysis of growth curves. Sankhya, A, 32, 53-64.

Generalized growth model. Rao's adjusted estimates and Bayesian justification. Estimating region. Prediction of future observations. Unadjusted estimator and its interpretation.

Geisser, S. (1970): See Enis, P. and Geisser, S., #294.

Geisser, S. (1970): See Kappenman, R.F., Geisser, S. and Antle, C.E., #573.

346. Geisser, S. (1971): A note on linear discriminants via idempotent matrices. Tech. Rep. #163, School Statist., U. Minnesota.

Multiple linear discriminants. Maximization of distance among mean vectors relative to a common covariance matrix. Idempotent matrices.

Geisser, S. (1971): See Enis, P. and Geisser, S., #295.

347. Geisser, S. and Kappenman, R.F. (1971): A posterior region for parallel profile differentials. Psychometrika, 36, 71-78.

Generalization of Geisser's solution for a single profile differential (#341). Bayes estimate and posterior region for vector of profile differentials. Application to repeated measurements experiments.

Geisser, S. (1972): See Lee, J.C. and Geisser, S., #717.

Geisser, S. (1972): See Lee, J.C. and Geisser, S., #718.

Geng, S. (1971): See Siotani, M., Chou, C. and Geng, S., #1040.

348. Ghosh, B.K. (1966): Asymptotic expansions for the moments of the distribution of correlation coefficient. Biometrika, 53, 258-262.

Asymptotic expansion of μ_1' and the central moments μ_2, μ_3, μ_4 of the noncentral distribution of r, the sample correlation coefficient. Expansion of hypergeometric function. Power series. Comparison of true values of μ_1', σ, β_1 and β_2 with approximate ones for various sample sizes and values of ρ.

349. Ghosh, M.N. (1963): Hotelling's generalized T^2 in the multivariate analysis of variance. J. Roy. Statist. Soc., B, 25, 358-367.

MANOVA based on Hotelling-Lawley statistic T^2. Power function of T^2-test: monotonic increasing function of each of the population characteristic roots separately. Simultaneous confidence intervals for all linear functions of the means. Expectation of T^2 for all p and variance for p = 3, 4. Rationale for splitting up the statistic into its components. Illustrated by a block design.

350. Ghosh, M.N. (1964): On the admissibility of some tests of MANOVA. Ann. Math. Statist., 35, 789-794.

Admissibility of (i) Hotelling-Lawley criterion based on sum of roots (ii) Roy's largest root criterion. Application of technique used by Stein for admissibility of T^2. Multivariate linear hypotheses.

351. Ghurye, S.G. and Olkin, I. (1962): A characterization of the multivariate normal distribution. Ann. Math. Statist., 33, 533-541.

A characterization of the MVN distribution along the lines of Skitovic. Replacement of scalar coefficients by matrices. General case. A functional equation of Skitovic and its application. Singularity of the coefficient matrices.

352. Ghurye, S.G. and Olkin, I. (1969): Unbiased estimation of some multivariate probability densities and related functions. Ann. Math. Statist., 40, 1261-1271.

UMVU estimators of p.d.f.'s: (a) Wishart density with (i) Σ known up to a scalar multiple, (ii) $\Sigma = \sigma^2[I(1-\rho)+\rho J]$, σ and ρ unknown, (iii) Σ completely unknown. (b) MVN density with mean unknown and Σ unknown or partially known. (c) MVN density with μ restricted and (i) Σ unknown, (ii) Σ known up to a scalar multiple, (iii) $\Sigma = \sigma^2[(1-\rho)I+\rho J]$. (d) $\log|\Sigma|$, $tr\ \Sigma_1\Sigma_2^{-1}$, $(\mu_1-\mu_2)'\Sigma^{-1}(\mu_1-\mu_2)$.

353. Ghurye, S.G. and Olkin, I. (1971): Identically distributed linear forms and the normal distribution. Tech. Rep. #58, Dept. Statist., Stanford U.

Survey of characterization of normal distribution in terms of the distribution of linear forms of i.i.d. random variables. Univariate and multivariate normality. Extension of Eaton's work in MV case: eliminate condition of symmetric matrices and introduce infinite linear form.

354. Gideon, R.A. and Gurland, J. (1972): A method of obtaining the bivariate normal probability over an arbitrary polygon. Tech. Rep. #304, Dept. Statist., U. Wisconsin.

Circular normal integral over an arbitrary angular region. Probability of $X \in P$, where P is arbitrary polygonal region and X is general BVN r.v. Approximations and tables of coefficients.

355. Gilbert, E.S. (1968): On discrimination using qualitative variables. J. Amer. Statist. Assoc., 63, 1399-1412.

Comparison of linear discriminant functions when variables are dichotomous. Fisher's linear discriminant function. Two functions based on logistic model. Function based on mutual independence of variables. Monte Carlo study. Relative merits of techniques judged by correlation of optimal function with linear function under study.

356. Gilbert, E.S. (1969): The effect of unequal variance-covariance matrices on Fisher's linear discriminant function. Biometrics, 25, 505-515.

Behaviour of Fisher's LDF when $\Sigma_1 \neq \Sigma_2$. Comparison based on $f_2(x)/f_1(x)$ when (i) $\Sigma_1 = \Sigma_2$ and (ii) $\Sigma_1 \neq \Sigma_2$. Probability of misclassification. Canonical form. Risk estimation. Monte Carlo study of case $\Sigma_1 = d\Sigma_2$, where d is a known constant.

357. Gilliland, D.C. (1962): Integral of the bivariate normal distribution over an offset circle. J. Amer. Statist. Assoc., 57, 758-768.

Determination of probability that a point (x,y) will lie within a circle of radius r and centered at origin when (x,y) has BVN distribution. Applications.

358. Giri, N.C. (1962): On a multivariate testing problem. Calcutta Statist. Assoc. Bull., 11, 55-60.

Test of $\mu_1 = 0$ vs. $\mu_2 = 0$ when μ_1 and μ_2 are partitions of μ, possibly overlapping. LR test. Distribution of LR statistic under null hypothesis.

359. Giri, N., Kiefer, J. and Stein, C. (1963): Minimax character of Hotelling's T^2-test in the simplest case. Ann. Math. Statist., 34, 1524-1535.

Test of hypothesis concerning mean vector. Hotelling's T^2 test of level α maximizes among all level α tests the minimum power on each of usual contours where T^2 test has constant power. Result proved for $p = 2$ and $N = 3$.

360. Giri, N.C. (1964): On the likelihood ratio test of a normal multivariate testing problem. Ann. Math. Statist., 35, 181-189. [Correction: 35 (1964), 1388].

Discriminant analysis with covariables. LR tests (conditional and unconditional): Uniformly most powerful similar invariant. Distribution of LR statistics.

361. Giri, N. and Kiefer, J. (1964): Local and asymptotic minimax properties of multivariate tests. Ann. Math. Statist., 35, 21-35.

Proofs of local minimax, type D, asymptotic minimax (as noncentrality tends to infinity) properties. Hotelling's T^2 is locally minimax, asymptotically minimax test. R^2 is locally minimax. Types D and E regions.

362. Giri, N. and Kiefer, J. (1964): Minimax character of the R^2-test in the simplest case. Ann. Math. Statist., 35, 1475-1490.

The test based on the multiple correlation coefficient R^2, for the case of $p = 3$ and $n = 3$ (or 4) is shown to maximize the minimum power on contours of constant power. Minimax procedures in multivariate analysis.

363. Giri, N.C. (1965): On the complex analogues of T^2- and R^2- tests. Ann. Math. Statist., 36, 664-670.

Complex MVN. Maximum likelihood estimation of μ and Σ. Independence of the sample mean and variance and their distributions. Tests of hypotheses concerning mean and multiple correlation coefficients. Distributions of T^2 and R^2 statistics. Properties of the corresponding tests.

364. Giri, N.C. (1965): On the likelihood ratio test of a normal multivariate testing problem - II. Ann. Math. Statist., 36, 1061-1065.

X is MVN with $E(X) = \mu$ and $V(X) = \Sigma$. Writing $\Gamma = \Sigma^{-1}\mu = [\Gamma_1, \Gamma_2, \Gamma_3]$, the LR test of H_0: $\Gamma_2 = \Gamma_3 = 0$ against H_1: $\Gamma_2 \neq 0$, $\Gamma_3 = 0$ is shown to be UMPI similar. It is also UMP similar among all tests with power depending only on the maximal invariant. Determination of maximal invariants of $\bar{X}$ and S, and of μ and Σ, under a group of transformations. Distributions of the maximal invariants under H_0 and H_1.

365. Giri, N.C. (1968): Locally and asymptotically minimax tests of a multivariate problem. Ann. Math. Statist., 39, 171-178.

Test of H_0: $\mu = 0$ against H_1: $\mu_1 = 0$, $\mu'\Sigma^{-1}\mu = \lambda$. The LR test is not locally or asymptotically minimax (with respect to λ). Determination of tests that are either locally or asymptotically minimax. Hotelling's test.

366. Giri, N.C. (1968): On tests of the equality of two covariance matrices. Ann. Math. Statist., 39, 275-277. [Correction: 39 (1968), 1764].

$\theta_1, \ldots, \theta_p$ are roots of $\Sigma_1^{-1}\Sigma_2$. Locally best invariant test of H_0: $\theta_1 = \ldots = \theta_p = 1$ against H_1: $\Sigma\theta_i > p$. Based on $trS_2(S_1+S_2)^{-1}$.

367. Giri, N.C. (1971): On the distribution of a complex multivariate statistic. Arch. Math., 22, 431-435.

Complex analogues of #368.

368. Giri, N.C. (1971): On the distribution of a multivariate statistic. Sankhya, A, 33, 207-210.

Statistics used for testing the mean and variance simultaneously. Non-null distribution.

Giri, N.C. (1971): See Behara, M. and Giri, N.C., #84.

369. Glahn, H.R. (1969): Some relationships derived from canonical correlation theory. Econometrica, 37, 252-256.

Interpretation of canonical variables in terms of the variance. Formulae for variation explained by such functions. Comments on Hooper's trace correlation.

370. Gleser, L.J. (1966): A note on the sphericity test. Ann. Math. Statist., 37, 464-467. [Correction: 39 (1968), 684].

LR test statistic for H_0: $\Sigma = \sigma^2 I$ *against* H_1: $\Sigma \neq \sigma^2 I$. *Canonical form of the distribution. Representation in terms of the test for homogeneity of variances. Unbiasedness of LR test. Asymptotic non-null distribution.*

371. Gleser, L.J. (1966): The comparison of multivariate tests of hypothesis by means of Bahadur efficiency. Sankhya, A, 28, 157-174.

Discussion of Bahadur efficiency and its modifications. Asymptotic distribution of some MV statistics. Sufficient conditions for Bahadur efficiency to be applicable. Rao's U statistic, Olkin and Shrikhande statistic, T^2*-like statistics. Optimility of Hotelling's* T^2 *vis-a-vis non-invariant tests. Tests on equality of variance matrices: Hotelling's* T_0^2, *LR, Roy and Gnanadesikan tests.*

372. Gleser, L.J. and Olkin, I. (1966): A k-sample regression model with covariance. Multivariate Anal.-I, [Proc. 1st Internat. Symp., Krishnaiah, ed.], 59-72.

k-sample extension of Cochran-Bliss-Rao problem. Tests for equality of regression matrices. MLE of the regression coefficients and variance matrix (assumed equal). Simultaneous confidence bounds on regression coefficients. Exact null and limiting non-null distributions of LR test-statistic.

373. Gleser, L.J. (1968): On testing a set of correlation coefficients for equality: some asymptotic results. Biometrika, 55, 513-517.

Expression of test statistics for H_0: $\rho_{ij} = \rho$ *(against unspecified alternative) as a composite quadratic form. Limiting null distribution of the quadratic form. Asymptotic probability of type I error. Similarity of the tests. Maximization of* α. *Invalidity of the Aitkin-Nelson-Reinfurt test.*

374. Gleser, L.J. and Olkin, I. (1969): Testing for equality of means, equality of variances, and equality of covariances under restrictions upon the parameter space. Ann. Inst. Statist. Math., 21, 33-48.

Tests of the restricted hypotheses (i) H_{01}: Means, variances and covariances are equal with the correlation coefficient being restricted to the interval $[\rho_0, 1]$, (ii) H_{02}: All the variances and covariances are equal with the correlation coefficient lying in $[\rho_0, 1]$. The MLE, moments of MLE, asymptotic distributions of the MLE are found under both sets of hypotheses. The LR tests and the asymptotic distribution of the criteria are also found.

375. Gleser, L.J. and Olkin, I. (1970): Linear models in multivariate analysis. Essays Prob. Statist., (Roy Volume, Bose et al, eds.), 267-292.

Canonical form of Potthoff-Roy model. MLE of the parameters and LR tests. Exact null and asymptotic non-null distributions of the criterion. Asymptotic distributions of MLE. Models related to the present model and tests of hypotheses. Rank and identifiability. Maximal invariants.

376. Gleser, L.J. and Olkin, I. (1972): Estimation for a regression model with an unknown covariance matrix. Proc. Sixth Berkeley Symp. Math. Statist. Prob., 1, 541-568.

Inferences on regression model with estimated covariance matrix. Estimation of β, Σ. Asymptotic normality and exact distributions of MLE. Derivation of the distributions. Confidence regions for β: (i) directly (ii) using ancilliary statistic. Comparisons of the confidence regions by volume considerations. Example. Tables.

377. Gleser, L.J. and Watson, G.S. (1972): Estimation of a linear transformation. Tech. Rep. #20 (Series 2), Dept. Statist., Princeton U.

MLE in multivariate models: X_i and Y_i are MVN with variance matrix $\sigma^2 I$ and $E(X_i) = \mu_i$ and $E(Y_i) = B\mu_i$. Estimation of B, μ_i and σ^2. Maximization of the likelihood. Properties of MLE. Consistency of MLE's. Least squares. Fisher's linear function approach. Applications.

378. Gnanadesikan, M. and Gupta, S.S. (1970): A selection procedure for multivariate normal distributions in terms of the generalized variances. Technometrics, 12, 103-117.

Generalized variance as a measure of dispersion. Selection of a subset of k p-variate normal populations containing population with smallest (largest) generalized variance. Approximations to distribution of generalized variace: simulation studies. Performance of procedure evaluated on basis of risk functions. Example of procedure.

Gnanadesikan, R. (1957): See Roy, S.N. and Gnanadesikan, R., #965.

Gnanadesikan, R. (1958): See Roy, S.N. and Gnanadesikan, R., #967.

379. Gnanadesikan, R. (1959): Equality of more than two variances and of more than two dispersion matrices against certain alternatives. Ann. Math. Statist., 30, 177-184. [Correction: 31 (1960), 227-228].

Choice of one variance matrix as a standard (unknown) and compare the other k matrices with it. Union intersection principle. Simultaneous confidence bounds for $c(\Sigma_i \Sigma_0^{-1})$.

Gnanadesikan, R. (1959): See Roy, S.N. and Gnanadesikan, R., #970.

Gnanadesikan, R. (1959): See Roy, S.N. and Gnanadesikan, R., #971.

Gnanadesikan, R. (1961): See Wilk, M.B. and Gnanadesikan, R., #1169.

Gnanadesikan, R. (1962): See Roy, S.N. and Gnanadesikan, R., #978.

Gnanadesikan, R. (1962): See Smith, H., Gnanadesikan, R. and Hughes, J.B.

380. Gnanadesikan, R. (1963): Multivariate statistical methods for analysis of experimental data. Indust. Qual. Control, 19, 22-26 and 31-32.

Survey of data analytic techniques (internal comparisons) for multiresponse experiments: MANOVA, principal component analysis, distance functions, "step-down" procedures. Graphical analysis using gamma probability plots. Applications.

Gnanadesikan, R. (1964): See Wilk, M.B. and Gnanadesikan, R., #1170.

381. Gnanadesikan, R. and Wilk, M.B. (1969): Data analytic methods in multivariate statistical analysis. Multivariate Anal.-II, [Proc. 2nd Internat. Symp., Krishnaiah, ed.], 593-636.

Philosophy of data analysis in multivariate case. Reduction of dimensionality: linear (principal components), nonmetric, nonlinear singularities (graphical analysis). Relationships or dependencies among variables: multivariate regression analysis, canonical analysis. Classification procedures: distance functions, cluster analysis, quality control. Overview of multivariate statistical models. Bibliography.

382. Gnanadesikan, R. and Lee, E.T. (1970): Graphical techniques for internal comparisons amongst equal degrees of freedom groupings in multiresponse experiments. Biometrika, 57, 229-237.

Data analytic techniques for multiresponse experiments. Simultaneous comparison of k dispersion matrices each based on ν degrees of freedom. "Size" of dispersions based on probability plotting techniques. "Size" discrepancies based on (i) arithmetic mean of eigenvalues, (ii) geometric mean of eigenvalues. Monte Carlo studies to obtain approximate null distributions for (i) and (ii). Applications: assessment of effects (main and interaction) in factorial experiments, comparison of k observed covariance matrices (MANOVA or multiple discriminant analyses).

Gnanadesikan, R. (1971): See Andrews, D.F., Gnanadesikan, R. and Warner, J.L., #43.

383. Gnanadesikan, R. (1972): Methods for evaluating similarity of marginal distributions. Statistica Neerlandica, 26, 69-78.

Assessment of degree of "commonality" of marginal distributions of components of a supposed MVN. Data analytic approach: probability plotting techniques and data-based transformations for improving normality.

384. Gnanadesikan, R. and Kettenring, J.R. (1972): Robust estimates, residuals, and outlier detection with multiresponse data. Biometrics, 28, 81-124.

Data-oriented techniques. Robust estimation of multivariate location and dispersion parameters. Analysis of multidimensional residuals (principal components and least squares). Detection of outliers. Probability plotting techniques. Internal comparisons. Monte Carlo studies.

Goldman, G.E. (1967): See Elashoff, J.D., Elashoff, R.M. and Goldman, G.E., #290.

385. Gollob, H.F. (1968): Confounding of sources of variation in factor-analytic techniques. Psychol. Bull., 70, 330-344.

Analysis of two-way classifications using a combination of analysis of variance and factor analysis (FANOVA). Decomposition of interaction parameter. Alternative to factor analysis of data matrix containing row and column main effects. Tukey's vacuum cleaner.

386. Gollob, H.F. (1968): A statistical model which contains features of factor analytic and analysis of variance techniques. Psychometrika, 33, 73-116.

Combining factor analysis and ANOVA in a two-way design (FANOVA). Decomposition of interaction parameter using factor analytic techniques. Interpretatons. Tests of significance.

387. Good, I.J. (1969): Some applications of the singular decomposition of a matrix. Technometrics, 11, 823-831.

Representation of a matrix M of rank r as a linear function of matrices of rank 1. Applications to solutions of equations, LS theory, principal components and eigenvalue problems.

388. Goodman, M.M. (1968): A measure of 'overall variability' in populations. Biometrics, 24, 189-192.

Use of generalized variance for measuring overall variabiliby in multivariate populations. Table of values of generalized variances for an example. Discussion.

389. Goodman, N.R. (1963): Statistical analysis based on a certain multivariate complex Gaussian distribution (an introduction). Ann. Math. Statist., 34, 152-177.

Introductory paper on complex analogues of MVN, Wishart, multiple correlation. Algebra of complex matrices and Hermitian forms. MLE of μ and Σ. Sufficiency. Characteristic function of complex Wishart and its trace. Transformations and Jacobians. Comparison between real and complex distribution theory. Applications to time series.

390. Goodman, N.R. (1963). The distribution of the determinant of a complex Wishart distributed matrix. Ann. Math. Statist., 34, 178-180.

Representation of the distribution of a complex Wishart determinant as the product of chi-squared densities. Comparison with real case.

391. Goodman, N.R. and Dubman, M.R. (1969): Theory of time-varying spectral analysis and complex Wishart matrix process. Multivariate Anal.-II, [Proc. 2nd Internat. Symp., Krishnaiah, ed.], 351-365.

Complex Gaussian and Wishart. Central and noncentral cases. Joint distribution of two correlated central complex Wisharts. Bivariate 2x2 complex Wishart distribution. Multivariate complex Wishart distribution.

392. Gordon, F.S. and Mathai, A.M. (1972): Characterizations of the multivariate normal distribution using regression properties. Ann. Math. Statist., 43, 205-229.

A characterization of MVN. A number of results on differentiation and integration of vector characteristic functions.

393. Govindarajulu, Z. (1967): Two sided confidence limits for $P(X < Y)$ based on normal samples of X and Y. Sankhya, B, 29, 35-40.

Expressions for the sample size needed to attain a specified confidence length and coefficient for $p = P\{X < Y\}$. Cases considered: (i) (X,Y) are BVN with unknown means but known covariance structure, (ii) independent (X,Y), (iii) X and Y are paired with variances and covariance unknown. Tables.

394. Gower, J.C. (1966): Some distance properties of latent root and vector methods used in multivariate analysis. Biometrika, 53, 325-338.

Q and R techniques. Analyses of distances between points. Representation by smaller dimensions. Principal component, factor, Mahalanobis' D^2 analyses. Qualitative variates (association) and quantitative relationships. Dualities. An extensive discussion of parametric and nonparametric techniques of distance studies.

395. Gower, J.C. (1966): A Q-technique for the calculation of canonical variates. Biometrika, 53, 588-590.

Canonical variates and their determination. Applications to other methods of distance analysis. Mahalanobis' distance function and its bias. Discriminant functions.

396. Gower, J.C. (1967): A comparison of some methods of cluster analysis. Biometrics, 23, 623-637.

A study of relative merits and performances of three methods: (i) Sokal and Michener, (ii) Edwards and Cavalli-Sforza, (iii) Williams and Lambert. Geometrical interpretation and clustering criteria. Modifications. Method of maximizing multiple correlations.

397. Gower, J.C. (1968): Adding a point to vector diagrams in multivariate analysis. Biometrika, 55, 582-585.

Discussion on the addition of a point in a multidimensional space. Determination of co-ordinates in terms of the axes of original set of points. Effects on the formulae in principal component analysis, cluster analysis and D^2. An example.

398. Graybill, F.A. and Milliken, G. (1969): Quadratic forms and idempotent matrices with random elements. Ann. Math. Statist., 40, 1430-1438.

Quadratic functions based MVN vectors. Distributions when the elements of the matrix are random. Wishartness of quadratic expressions. Characteristic function techniques.

Graybill, F.A. (1970): See Kingman, A. and Graybill, F.A., #624.

399. Greenberg, B.G. and Sarhan, A.E. (1959): Matrix inversion, its interest and application in analysis of data. J. Amer. Statist. Assoc., 54, 755-766.

Inversion of different types of patterned matrices. Applications to different situations.

400. Greenberg, B.G. and Sarhan, A.E. (1960): Generalization of some results for inversion of patterned matrices. Contrib. Prob. Statist., (Hotelling Volume, Olkin et al, eds.), 216-223.

Generalizations of Roy-Sarhan paper on inversion of patterned matrices. Several special cases.

Greenberg, B.G. (1960): See Roy, S.N., Greenberg, B.G. and Sarhan, A.E., #975.

Greenhouse, S.W. (1958): See Geisser, S. and Greenhouse, S.W., #333.

401. Greenhouse, S.W. and Geisser, S. (1959): On methods in the analysis of profile data. Psychometrika, 24, 95-112.

Profile analysis. Mixed model. Approximate F-tests when covariance matrix is arbitrary. Hypotheses: equality of test means, equality of group means, no group x test interaction (parallel profiles). Exact procedures based on MANOVA. Example illustrating exact and approximate procedures. Approximate tests when covariance matrices for g groups are not equal.

Grizzle, J.E. (1962): See Elston, R.C. and Grizzle, J.E., #293.

Grizzle, J.E. (1966): See Cole, J.W.L. and Grizzle, J.E., #182.

402. Grizzle, J.E. and Allen, D.M. (1969): Analysis of growth and dose response curves. Biometrics, 25, 357-381.

Analysis of longitudinal data: growth curves, dose response curves over time. Estimation and tests of hypothesis from view point of analysis of covariance (Rao). Multivariate analogue of weighted and unweighted least squares. Use of all, none or a subset of covariables. Tests of multivariate linear hypotheses. Confidence bounds for response curve. Techniques exemplified by several sets of data.

403. Grizzle, J.E. (1970): An example of the analysis of a series of response curves and an application of multivariate multiple comparisons. Essays Prob. Statist., (Roy Volume, Bose et al, eds.), 311-326.

Series of observations made over time or over a succession of doses: response curves. Reduction of multivariate data. Estimation of functional relationship for each individual. Analysis of estimated parameters using MANOVA techniques. Multiple comparisons. Illustration of techniques with a numerical example.

404. Groves, T. and Rothenberg, T. (1969): A note on the expected value of an inverse matrix. Biometrika, 56, 690-691.

Property of expectation of inverse matrix: $E(A^{-1})-\{E(A)\}^{-1}\geq 0$ *with A a real, symmetric, positive definite matrix. Convex function.*

405. Guenther, W.C. (1961): Circular probability problems. Amer. Math. Monthly, 68, 541-544.

Integrals associated with uncorrelated BVN. Probability of covering randomly selected point: (i) in a circle, (ii) on the circumference of a circle.

406. Guenther, W.C. (1961): On the probability of capturing a randomly selected point in three dimensions. SIAM Rev., 3. 247-251.

Coverage problems in three dimensions: X, Y, Z are independently N(0,1). Areas and volumes of a sphere. Extension of #405.

407. Guenther, W.C. (1964): A generalization of the integral of the circular coverage function. Amer. Math. Monthly, 71, 278-283.

Coverage problem in n dimensions: X_i are independently N(0,1) for $i = 1,\ldots,n$. Areas and volumes. Integral expressed in terms of integrals of noncentral χ^2. Spherical co-ordinates. Extension of #405 and #406.

408. Guenther, W.C. and Terragno, P.J. (1964): A review of the literature on a class of coverage problems. Ann. Math. Statist., 35, 232-260.

A review paper discussing multiple integration over offset regions involving normal weight functions. Probability contents in multidimensional case of a domain D. Bibliographic.

409. Gumbel, E.J. (1960): Multivariate distributions with given margins and analytical examples. Bull. Inst. Internat. Statist., 37, 363-373.

A general discussion of the properties of bivariate distributions. Existence of bivariate densities when marginals exist. Multidimensional extensions. Symmetry and independence. Examples. Bivariate normal.

410. Gumbel, E.J. (1960): Bivariate exponential distributions. J. Amer. Statist. Assoc., 55, 698-707.

Two types of bivariate exponentials. Properties of marginal, conditional distributions. Correlations.

411. Gumbel, E.J. (1961): Bivariate logistic distributions. J. Amer. Statist. Assoc., 56, 335-349.

Two types of bivariate logistic introduced. A study of the properties of the first one. Regression and conditional distributions. Comparison with normal.

Gupta, A.K. (1967): See Pillai, K.C.S. and Gupta, A.K., #886.

Gupta, A.K. (1968): See Pillai, K.C.S. and Gupta, A.K., #889.

Gupta, A.K. (1969): See Pillai, K.C.S. and Gupta, A.K., #892.

412. Gupta, A.K. (1971): Distribution of Wilks' likelihood-ratio criterion in the complex case. Ann. Inst. Statist. Math., 23, 77-87.

Test for H_0: $\mu = 0$ in CMVN case when μ is the matrix of means. Wilks' LR criterion. Null distribution for any dimensions and degrees of freedom. Distribution function. Closed forms for $p = 2, 3$. Tables of correction to χ^2.

413. Gupta, A.K. (1971): Noncentral distribution of Wilks' statistic in MANOVA. Ann. Math. Statist., 42, 1254-1261.

Distribution of Wilks' Λ in the noncentral linear case (i.e., rank of noncentrality parameter is 1). Expression of the criterion as the product of independent random variables. Exact distribution (non-null) for $p = 2(1)5$. General case. Examination of Posten-Bargmann approximation.

414. Gupta, R.P. (1964): Some extensions of the Wishart and multivariate t-distributions in the complex case. J. Indian Statist. Assoc., 2, 131-136.

Derivation of the complex Wishart distribution by the method of random orthogonal transformations Null and non-null (linear case). Definition of complex analogues of Dunnet-Sobel and Cornish multivariate t-distributions.

415. Gupta, R.P. (1965): Asymptotic theory for principal component analysis in the complex case. J. Indian Statist. Assoc., 3, 97-106.

The asymptotic distributions of the roots and vectors as applied to complex normal case. Derivation using Anderson's method. Extension of Kshirasagar's test for goodness-of-fit of a single non-isotropic principal component to complex case.

Gupta, R.P. (1965): See Kshirasagar, A.M. and Gupta, R.P., #678.

Gupta, R.P. (1965): See Kshirasagar, A.M. and Gupta, R.P., #679.

416. Gupta, R.P. (1967): Latent roots and vectors of a Wishart matrix. Ann. Inst. Statist. Math., 19, 157-165.

LR tests of hypotheses concerning latent roots and vectors of Σ. Asymptotic distributions in null case. Derivation of MLE of the parameters under various hypotheses. Alternative proof for independence and Wishartness of $S_{11.2}$ and $S_{12}S_{22}^{-1}S_{21}$.

417. Gupta, R.P. and Kabe, D.G. (1970): The noncentral multivariate Beta distribution in planar case. Trabajos Estadist., 21, 61-67.

Development of distribution of matrix V, where A = CVC' and A+B = CC' with A, B independent and Wishart. Noncentral multivariate beta distribution.

418. Gupta, R.P. and Kabe, D.G. (1971): Distribution of certain factors useful in discriminant analysis. Ann. Inst. Statist. Math., 23, 97-103.

Testing for direction and collinearity using generalization of factorization given by Bartlett. Alternative expressions for factors in terms of elements of T instead of elements of L as obtained by Radcliffe (#921).

Gupta, R.P. (1972): See Kabe, D.G. and Gupta, R.P., #570.

419. Gupta, S.S. (1963): Probability integrals of multivariate normal and multivariate t. Ann. Math. Statist., 34, 792-828.

Survey of work on multivariate normal probability integral and related functions. Multivariate analogue of Student's t distribution. Tables provided for probability integral in equicorrelated case.

420. Gupta, S.S. (1963): Bibliography on the multivariate normal integrals and related topics. Ann. Math. Statist., 34, 829-838.

Bibliography on some 200 articles. Classified into 12 sections. Cross-classification.

421. Gupta, S.S., Pillai, K.C.S. and Steck, G.P. (1964): On the distribution of linear functions and ratios of linear functions of ordered correlated normal random variables with emphasis on range. Biometrika, 51, 143-151.

Distribution of range: equally correlated random variables and general case. General case: distribution of range obtained for sample sizes 2, 3, 4. Equally correlated case: expressions for probability integral, percentage points and moments of linear functions in terms of corresponding expressions for uncorrelated case. Distribution of ratio of certain linear functions.

422. Gupta, S.S. and Pillai, K.C.S. (1965): On linear functions of ordered correlated normal random variables. Biometrika, 52, 367-379.

Continuation of #42. Characteristic functions of order statistics and linear functions of them for BVN and TVN with general correlation matrix. Formulae for mean, variance and covariance of order statistics. First and second moments for linear function of the order statistics. Joint distribution of range and mid-range for TVN case: distributions of mid-range/range ratio. BLUE of common mean of TVN. Applications to life testing and time series.

423. Gupta, S.S. (1966): On some selection and ranking procedures for multivariate normal populations using distance functions. Multivariate Anal.-I, [Proc. 1st Internat. Symp., Krishnaiah, ed.], 457-475.

General discussion. Procedures for selecting the subset to contain MV population with the largest λ_i, the Mahalanobis distance with variance known. Simultaneous cnnfidence bounds on $(p+\lambda_i)/(p+\lambda_i)$. Smallest λ. Probability of incorrect selection. Distributions of some statistics involved. Tables.

424. Gupta, S.S. and Panchapakesan, S. (1969): Some selection and ranking procedures for multivariate normal populations. Multivariate Anal.-II, [Proc. 2nd Internat. Symp., Krishnaiah, ed.], 475-505.

Selection and ranking based on (i) Multiple correlation coefficient: conditional and unconditional cases. Asymptotic results and procedures based on a transformation of the multiple correlation coefficient. Problem of selecting a subset containing the population associated with smallest population multiple correlation coefficient. (ii) Conditional generalized variance, in the unconditional case. Tables.

Gupta, S.S. (1970): See Gnanadesikan, M. and Gupta, S.S., #378.

425. Gupta, S.S. and Studden, W.J. (1970): On some selection and ranking procedures with applications to multivariate populations. Essays Prob. Statist. (Roy Volume, Bose et al, eds.), 327-338.

Probabilities of correct selection and their infima Selection and ranking based on Mahalanobis distance with Σ known and unequal or Σ unknown and unequal.

426. Gurland, J. (1968): A relatively simple form of the distribution of the multiple correlation coefficient. J. Roy. Statist. Soc., B, 30, 276-283.

MLE of multiple correlation coefficient. The noncentral case. Characteristic function represented as that of the ratio of two independent random variables. Distribution of $u = R^2/(1-R^2)$. Expansions for the c.d.f. Approximations.

Gurland, J. (1969): See Mehta, J.S. and Gurland, J., #787.

Gurland, J. (1969): See Mehta, J.S. and Gurland, J., #788.

427. Gurland, J. and Milton, R.C. (1970): Further consideration of the distribution of the multiple correlation coefficient. J. Roy. Statist. Soc., B, 32, 381-394.

General expansion of the characteristic function of #426 in an infinite series. Expression of the c.d.f. of the multiple correlation coefficient in an infinite series involving the c.d.f. of F-distribution. Convergence behaviour of the expansions. Approximations to the distribution of multiple correlation coefficient. Comparisons of Gurland's approximation with Fisher's Z. Tables.

Gurland, J. (1972): See Gideon, R.A. and Gurland, J., #354.

428. Guttman, I. (1970): Construction of β-content tolerance regions at confidence level γ for large samples from the k-variate normal distribution. Ann. Math. Statist., 41, 376-400.

Construction of tolerance region in MVN case. Coverage and its moments to $O(n^{-1})$. Approximations to the distribution of coverage by incomplete Beta. Tolerance region of β-content and pre-assigned expectation. Efficiencies. Tables of constants. Differentiation formulae.

429. Guttman, I. (1971): A Bayesian analogue of Paulson's lemma and its use in tolerance region construction when sampling from a multivariate normal. Ann. Inst. Statist. Math., 23, 67-76.

Multivariate generalization of Paulson's lemma relating the statistical tolerance region with a prediction region. Bayesian approach. Predictive β-confidence region. Relation to posterior expectation of coverage. Applications to MVN. Ignorance prior of Geisser. Predictive density.

Guttman, I. (1971): See Tan, W.Y. and Guttman, I., #1121.

Hagis, P., Jr. (1963): See Krishnaiah, P.R., Hagis, P., Jr. and Steinberg, L., #641.

Hagis, P., Jr. (1965): See Krishnaiah, P.R., Hagis, P., Jr. and Steinberg, L., #648.

430. Hahn, G.J. (1969): Factors for calculating two-sided prediction intervals for samples from a normal distribution. J. Amer. Statist. Assoc., 64, 878-888.

Application of the MV t distribution to set up prediction interval of prescribed confidence for k future observations from a normal distribution. Approximations. Comparisons. Mathematical formulae for the computations. Example. Tables and graphs.

431. Hahn, G.J. (1970): Additional factors for calculating prediction intervals for samples from a normal distribution. J. Amer. Statist. Assoc., 65, 1668-1676.

A continuation of #430. Extended tables to include one-sided prediction intervals. Computation and comparison with existing tables of Dunnett, Steffens, Krishnaiah and Armitage, Dunn and Massey. Tables.

432. Hahn, G.J. and Hendrickson, R.W. (1971): A table of percentage points of the distribution of the largest absolute value of k Student t variates and its applications. Biometrika, 58, 323-332.

Percentage points of the maximum value of $|t|$ when sampling from a k-variate Student's t distribution with ν degrees of freedom and common correlation coefficient ρ. A number of combinations of the parameters is considered. Several applications of the tables are listed: (i) Prediction intervals to contain all of k future means when σ^2 is estimated by pooling, (ii) Multiple comparison between k treatment means and a control mean, (iii) SCI to contain all of k population means, (iv) SCI to contain k regression coefficients.

Haller, H.S. (1969): See Bell, C.B. and Haller, H.S., #85.

433. Halperin, M. and Mantel, N. (1963): Interval estimation of non-linear parametric functions. J. Amer. Statist. Assoc., 58, 611-627.

Interval estimate of a nonlinear function $f(\mu_1,\mu_2)$ when X is BVN with unknown μ_1, μ_2 and known Σ. Bounds are function of χ^2. Comparison with "delta method".

434. Halperin, M. (1964): Interval estimation of non-linear parametric functions, II. J. Amer. Statist. Assoc., 59, 168-181.

Extension of #433. Interval estimates for $f(\mu_1, \mu_2, \ldots, \mu_k)$ possibly non-linear. Based on k sample means with unknown expectation μ_i, known variance σ_i^2/n_i and arbitrary correlation coefficient. Asymptotic 1-α regions. Ellipsoidal and rectangular constraints. Detailed results for bivariate case.

435. Halperin, M. (1967): An inequality on a bivariate Student's 't' distribution. J. Amer. Statist. Assoc., 62, 603-606.

Inequality useful in multiple comparisons: $Pr(t_i^2 \le C_i^2) > \Pi Pr(t_i^2 < C_i^2)$ for i = 1, 2 and C_i any real numbers. Joint distribution of t_i's: bivariate Student's t. Analogue of Sidak's results.

436. Halperin, M. (1967): A generalization of Fieller's theorem to the ratio of complex parameters. J. Roy. Statist. Soc., B, 29, 126-131.

Derivation of complex analogue of Fieller's theorem. Confidence regions for real and imaginary parts of ratio of complex parameters. Exact confidence region (assuming normality) based on Hotelling's T^2. Approximate region. Joint distribution of two correlated t variates.

Hamdan, M.A. (1971): See Reiger, M.H. and Hamdan, M.A., #952.

437. Han, C.P. (1968): Testing the homogeneity of a set of correlated variances. Biometrika, 55, 317-326.

Testing equality of k variances when X is MVN and correlation coefficients are equal and unknown. Comparison of four procedures: (i) likelihood ratio, (ii) modified Bartlett test, (iii) multiple correlation test, (iv) F(max) test. Monte Carlo studies. Adequacy of asymptotic results for p = 3 and 4.

438. Han, C.P. (1968): A note on discrimination in the case of unequal covariance matrices. Biometrika, 55, 586-587.

Extension of Bartlett and Please: mean difference is not zero. Discrimination using likelihood ratio procedure. $\Sigma_1 \neq \Sigma_2$ but correlation coefficients are equal. Components of discriminant function: size, shape and sum of square of variates.

439. Han, C.P. (1969): Distribution of discriminant function when covariance matrices are proportional. Ann. Math. Statist., 40, 979-985.

Discriminant function with $\Sigma_2 = \sigma^2\Sigma_1$ (σ^2 and Σ_1 known). Likelihood ratio procedures. Distribution of U, the discriminant function: (i) μ_1 and μ_2 known and equal, U is χ^2, (ii) μ_1 and μ_2 known and unequal, U is noncentral χ^2, (iii) μ_1 and μ_2 unknown, asymptotic distribution for U. Techniques for asymptotic distribution: "studentization", expansion and inversion of characteristic function.

440. Han, C.P. (1970): Distribution of discriminant function in circular models. Ann. Inst. Statist. Math., 22, 117-125.

Discriminant functions when Σ_1 and Σ_2 are circular. Transformation of covariance matrices to canonical form. Distribution of discriminant function: (i) all parameters known, (ii) covariance matrices are unknown and mean vectors are known, (iii) all parameter unknown Asymptotic expansion for distribution. "Studentization" technique. Inversion of characteristic function.

441. Hannan, E.J. (1961): The general theory of canonical correlation and its relation to functional analysis. J. Austral. Math. Soc., 2, 229-242.

Generalization of Lancaster's results by spectral decomposition of operators on a Hilbert space. Relationship to stationary Gaussian processes. Spectral theoretic presentation.

442. Hannan, E.J. (1965): Group representations and applied probability. J. Appl. Prob., 2, 1-68.

Section 8.2 applies to multivariate analysis. Geometric interpretation of transformations. Work of James using the properties of the orthogonal group.

443. Hannan, J.F. and Tate, R.F. (1965): Estimation of the parameters for a multivariate normal distribution when one variable is dichotomized. Biometrika, 52, 664-668.

MVN model with one of the component variables observable only in dichotomized form. Unknown point of dichotomy. Estimation of (i) vector of correlations between components of X and dichotomous variable (ii) multiple correlation coefficient (iii) point of dichotomy. Biserial estimators. ML estimators. Asymptotic variances.

Hanson, D.L. (1964): See Owen, D.B., Craswell, K.J. and Hanson, D.L., #861.

444. Hanumara, R.C. and Thompson, W.A., Jr. (1968): Percentage points of the extreme roots of a Wishart matrix. Biometrika, 55, 505-512.

Approximate percentage points of distributions of largest and smallest characteristic roots of a Wishart matrix with $\Sigma = I$. Tables for $\alpha = 0.005, 0.01, 0.025, 0.05$ and $p = 2(1)10$. Pillai's approximation. Application to simultaneous confidence intervals for variance components in two-way (mixed) classification with unequal variances.

Haq, M.S. (1969): See Fraser, D.A.S. and Haq, M.S., #311.

445. Harley, B.I. (1957): Relation between the distributions of non-central t and of a transformed correlation coefficient. Biometrika, 44, 219-224.

Approximation for distribution of $r\sqrt{n-2}/(1-r^2)^{\frac{1}{2}}$ when $\rho \neq 0$. Comparison of moments of $r/(1-r^2)^{\frac{1}{2}}$ when $\rho \neq 0$ with moments of a noncentral t variate.

446. Harley, B.I. (1957): Further properties of an angular transformation of the correlation coefficient. Biometrika, 44, 273-275.

An alter ative way of establishing that $\sin^{-1} r$ is unbiased for arcsine ρ in the BVN CASE. Achieved by considering a transformation of the standardized BVN variates (U, V) to the variates (X, Y), where $X = (U-\rho V)/(1-\rho^2)^{\frac{1}{2}}$ and $Y = V$. X and Y are independent.

447. Harrison, P.J. (1968): A method of cluster analysis and some applications. Appl. Statist., 17, 226-236.

Analysis of dichotomous data. Quantitative and multistate variables. Programs for cluster analysis and classificatory problems. Imperial Chemical Industries routines and their applications.

448. Harter, H.L. (1960): Circular error probabilities. J. Amer. Statist. Assoc., 55, 723-731.

(X, Y) independent normal with means zero and variances σ_X^2 and σ_Y^2 $(\sigma_X^2 \geq \sigma_Y^2)$. Representation in an integral form probability that (X, Y) lies in a circle centered at $(0, 0)$ and of radius $K\sigma_X$. Computation of integral by trapezoidal rule. Tables of values of K for given probability.

449. Hartley, H.O. and Hocking, R.R. (1971): The analysis of incomplete data. Biometrics, 27, 783-823.

MVN distribution. Estimation of μ and Σ with a general type of incompleteness. Anderson-Bargmann-Trawinski models. Likelihood procedures or methods dependent on such procedures. Nested problems. Censored and grouped data. Confidence regions based on such data. Univariate problems of ANOVA and regression analysis. Examples. Convergence of iterative solutions. Significant discussion by Woodbury, Fienberg, Rubin. Reply to the discussion. Mostly on MVN case.

450. Hashiguchi, S. and Morishima, H. (1969): Estimation of genetic contribution of principal components to individual variates concerned. Biometrics, 25, 9-15.

Latent vectors and roots of phenotypic (observed) correlation matrix. Additivity of genetic and environmental factors. Genetic fraction of the latent root. Regression coefficient vector of the observation on the principal component. Genetic vector. Genetic contribution of the principal component. Example. Relationship of genetic vector to the component vector of the genetic correlation matrix.

451. Hayakawa, T. and Kabe, D.G. (1965): On testing the hypothesis that submatrices of the multivariate regression matrices of k populations are equal. Ann. Inst. Statist. Math., 17, 67-73.

$Y' = [X', Z']$ is MVN with $E(X) = BZ$ where $B' = (B_1', B_2')$. The hypothesis under test is H_0: $B_{11} = B_{21} = \ldots = B_{k1} = B_1^0$ for k sample case. Here B_1^0 is unspecified. MLE of the parameters B and of Var $X = \Sigma$. Null distribution of the LR criterion. Distribution of MLE of Σ when H_0 is true and B_1^0 is estimated.

452. Hayakawa, T. (1966): On the distribution of a quadratic form in a multivariate normal sample. Ann. Inst. Statist. Math., 18, 191-201.

Zonal polynomials. Distribution of a positive definite quadratic form in matrix variates. Decomposition of quadratic forms. Wishart and pseudo-Wishart matrices. $P\{X'AX \leq \Omega\}$. Joint density of the roots of the determinantal equation $|X'AX-\lambda I| = 0$. Distribution of the trace of $X'AX$. If W is Wishart, then the distribution of $R = (X'AX)^{\frac{1}{2}}[X'AX\ W]^{-1}(X'AX)^{\frac{1}{2}}$ and of the roots of R. This paper is related to Khatri (#601).

453. Hayakawa, T. (1967): On the distribution of the maximum latent root of a positive definite symmetric random matrix. Ann. Inst. Statist. Math., 19, 1-17. [Correction: 21 (1969), 221].

Zonal polynomials and some expansions related to them. Transformations and Jacobians involved in the transformations. An incomplete integral involving a zonal polynomial. Density functions of, (i) maximum latent root of the quadratic form X'AX, (ii) maximum l.r. of a noncentral Wishart matrix with unit covariance, (iii) maximum l.r. of a noncentral Beta-matrix, (iv) maximum canonical correlation coefficient in the non-null case. All based on direct integrations and transformations. Table of values of zonal polynomials of direct sums of matrices.

454. Hayakawa, T. (1969): On the distribution of the latent roots of a positive definite random symmetric matrix I. Ann. Inst. Statist. Math., 21, 1-21.

Orthogonal polynomials of matrix argument. Hermite polynomials, Laguerre polynomials and generating functions. Jacobians of transformations. Maximum latent root, integrals. Noncentral Wishart, p.d.f. expressed in Hermite polynomials. Maximum latent root as a series in Hermite polynomials. Trace of a noncentral Wishart. A new class of polynomials. Noncentral MV quadratic form, its maximum root and trace. All p.d.f.'s expressed in Hermite polynomials.

455. Haynsworth, E.V. (1961): Special types of partitioned matrices. J. Res. Nat. Bur. Standards, B, 65, 7-12.

Some new results on certain types of partitioned matrices.

Healy, M.J.R. (1957): See Ashton, E.H., Healy, M.J.R. and Lipton, S., #51.

456. Healy, M.J.R. (1965): Computing a discriminant function from within-sample dispersions. Biometrics, 21, 1011-1012.

A discussion of numerical aspects of discriminant function analysis.

457. Healy, M.J.R. (1968): Multivariate normal plotting. Appl. Statist., 17, 157-161.

Data analytic techniques for detecting outliers in MVN distribution. Quadratic form of the p.d.f. and χ^2. Normality and its tests. Plotting on normal graph paper. Transformation to Gamma variates.

458. Healy, M.J.R. (1969): Rao's paradox concerning multivariate tests of significance. Biometrics, 25, 411-413.

Reason behind a paradox: $\bar{X}$ is significant, $\bar{Y}$ is significant in BVN but $(\bar{X},\bar{Y})$ jointly tested at the same level are not significant. Geometrical display of the regions for which this possibility is valid. Univariate tests that are more powerful than bivariate ones.

459. Heck, D.L. (1960): Charts of some upper percentage points of the distribution of the largest characteristic root. Ann. Math. Statist., 31, 625-642.

Roy's largest root criterion. Upper percentage points of largest root. Use of Pillai's approximation for the cumulative distribution function of the largest root. Charts for variety of combinations of the parameters.

Hendrickson, R.W. (1971): See Hahn, G.J. and Hendrickson, R.W., #432.

460. Herr, D.G. (1967): Asymptotically optimal tests for multivariate normal distributions. Ann. Math. Statist., 38, 1829-1844.

Hoeffding's results on LR tests. Extension to distributions other than multinomial. Probabilities of large deviations in MVN. LR procedures. Example of a LR test better than Roy's test at most alternatives.

461. Hewett, J.E. and Bulgren, W.G. (1971): Inequalities for some multivariate f-distributions with applications. Technometrics, 13, 397-402.

An inequality on a multivariate f-distribution (i.e., jointly distributed f-variates). Krishnaiah's and Jensen's bivariate f-distributions. Applications to simultaneous prediction intervals.

462. Higgins, G.F. (1970): A discriminant analysis of employment in defense and nondefense industries. J. Amer. Statist. Assoc., 65, 613-622.

An application of discriminant analysis technique to study two types of industries. Discussion.

Higgins, J.J. (1972): See Bemis, B.M., Bain, L.J. and Higgins, J.J., #86.

463. Hill, G.W. and Davis, A.W. (1968): Generalized asymptotic expansions of Cornish-Fisher type. Ann. Math. Statist., 39, 1264-1273.

Cornish-Fisher expansions in terms of general distribution function that is analytic. Solution of the equation $F_n(x) = \Phi(u)$, for u in terms of x. Inverse expansion, percentage points. Examples for (i) Wilks' LR criterion for testing H_0: $\mu = 0$ (MANOVA), using Rao's expansion, (ii) Hotelling's T_0^2, using Ito's expansion. Expansin for $F_n(x)$ in terms of cumulants. Cornish-Fisher series. Some properties of the polynomials in the Cornish-Fisher expansions. Tables.

464. Hills, M. (1966): Allocation rules and their error rates. J. Roy. Statist. Soc., B, 28, 1-31.

General principles of allocation and their error rates. Inequality on the average apparent, optimum and average actual error. Estimation of the average actual error rate by ML and approximately unbiased estimation. Discussion of Lachenbruch and Cochran-Hopkins techniques. Special distributions including MVN. A discussion of Hill's results by Smith, Kerridge, Wagle, Plackett, Armitage and Cochran.

465. Hills, M. (1969): On looking at large correlation matrices. Biometrika, 56, 249-253.

Graphical methods for relating variables by plotting of correlation coefficients. Half-normal plotting and visual techniques. Example.

466. Hinkley, D.V. (1969): On the ratio of two correlated normal random variables. Biometrika, 56, 635-639. [Correction: 57 (1970), 683].

Distribution $W = X_1/X_2$ with X_1, X_2 BVN. Approximate distribution of W. Applications: (i) ratio $(-\alpha/\beta)$ in regression analysis, (ii) $(\alpha_1-\alpha_2)/(\beta_2-\beta_1)$, intersection of two regression lines.

467. Hinkley, D.V. (1972): Time-ordered classification. Biometrika, 59, 509-523.

Classification when after a certain time observations no longer come from the initial population considered. General discriminant statistics under assumptions of MVN. Marginal log-likelihoods. Estimation and confidence regions for time-point of shift. Power-to-size ratio. Efficiency.

468. Hocking, R.R. and Smith, W.B. (1968): Estimation of paramters in the multivariate normal distribution with missing observations. J. Amer. Statist. Assoc., 63, 159-173.

Estimation of parameters of MVN when missing observations are not restricted to a particular pattern. Estimation techniques: (i) subdivide data according to missing variates, (ii) initial estimates from data with no missing variates, (iii) modify initial estimates by adjoining optimally remaining groups in a sequential fashion. Consistency and asymptotic efficiency. ML estimators. Simulation studies.

Hocking, R.R. (1971): See Hartley, H.O. and Hocking, R.R., #449.

Hocking, R.R. (1972): See Smith, W.B. and Hocking, R.R., #1048.

Hodge, R.W. (1971): See Klatzky, S.R. and Hodge, R.W., #625.

469. Hogg, R.V. (1963): On the independence of certain Wishart matrices. Ann. Math. Statist., 34, 935-939.

MV generalizations of several important results valid in the univariate case. These are (i) Wishart distribution and its decomposition into k Wisharts, (ii) NSC for (B-A) to be positive definite, when A, B are idempotent, (iii) simple proof for Anderson's theorem #4.3.2, (iv) NSC for the forms X'B and X'AX to be independent.

Holland, D.A. (1960): See Pearce, S.C. and Holland, D.A., #868.

470. Holloway, L.N. and Dunn, O.J. (1967): The robustness of Hotelling's T^2. J. Amer. Statist. Assoc., 62, 124-136.

Robustness studies carried out for two-sample T^2 test when the variance matrices are unequal for $\alpha = .01$, .05. A number of combinations of parameters are considered. Departures from $\Sigma_1 = \Sigma_2$ are defined in terms of roots of $\Sigma_1\Sigma_2^{-1}$ all being equal to d. Graphs and charts are presented for level of significance and power of the test.

Honigfeld, G. (1969): See Feldman, S., Klein, D.F. and Honigfeld, G., #297.

471. Hooper, J.W. (1958): The sampling variance of correlation coefficients under assumptions of fixed and mixed variates. Biometrika, 45, 471-475.

Canonical correlations when some of the variates are random and some are fixed. Asymptotic variances and covariances are derived. Special case of total and multiple correlation coefficients.

472. Hooper, J.W. (1959): Simultaneous equations and canonical correlation theory. Econometrica, 27, 245-256.

Generalizations of the notion of correlation. Vector and trace correlations. Asymptotic sampling variances.

473. Hooper, J.W. and Zellner, A. (1961): The error of forecast for multivariate regression models. Econometrica, 29, 544-555.

MV linear model. Error of forecast. Hotelling's T^2. Forecast regions based on T^2. Example.

474. Hooper, J.W. (1962): Partial trace correlations. Econometrica, 30, 324-331.

Measure of trace correlation applied to partial regression situation. Asymptotic properties.

Hopkins, C.E. (1961): See Cochran, W.G. and Hopkins, C.E., #174.

475. Hopkins, J.W. and Clay, P.P.F. (1963): Some empirical distributions of bivariate T^2 and homoscedasticity criterion M under unequal variance and leptokurtosis. J. Amer. Statist. Assoc., 58, 1048-1053.

Monte Carlo studies on the robustness (under H_0) of two sample T^2 test and Bartlett's M-statistic against unequal variances, unequal sample sizes and non-normality defined by leptokurtosis. Tables.

476. Hopkins, J.W. (1966): Some considerations in multivariate allometry. Biometrics, 22, 747-760.

Multivariate population covariance matrix $\Sigma = \psi + \Delta$, with ψ being systematic and Δ being random components. Growth and size patterns. Principal components. Tests for goodness of fit. Application to interrelationships among growth of nine body parts of rats.

477. Horst, P. (1961): Relations among m sets of measures. Psychometrika, 26, 129-149.

Extension of Hotelling's canonical correlation analysis to more than two sets of variables. Maximizing correlation.

478. Horton, I.F., Russell, J.S. and Moore, A.W. (1968): Multivariate-covariance and canonical analysis: a method for selecting the most effective discriminators in a multivariate situation. Biometrics, 24, 845-857.

Selection of a subset of variables. Conbination of techniques: multivariate covariance analysis and canonical correlation analysis. Application to analysis of gilgaied soil data.

479. Hudimoto, H. (1957): A note on the probability of the correct classification when the distributions are not specified. Ann. Inst. Statist. Math., 9, 31-36.

Modification of Hudimoto's earlier paper. Nonparametric classificatory procedures. Lower bounds on a probability associated with a distribution function.

480. Hughes, D.T. and Saw, J.G. (1970): Moments of a random orthogonal matrix. Tech. Rep. #21, Dept. Statist., U. Florida.

Evaluation of moments of a random orthogonal matrix. Orthospherical integrals. Approximations to expectation of $\exp\{-tr(A\phi B\phi')/2\}$. *Moments of* $tr(AHBH')$.

481. Hughes, D.T. and Saw, J.G. (1970): Distribution of Hotelling's generalized T_0^2. Tech. Rep. #22, Dept. Statist., U. Florida.

Moments of T_0^2 *in the non-null case and an approximation to the distribution. Power of* T_0 *test. Monotonicity. Proof alternative to Das Gupta et al (#217). Tables.*

482. Hughes, D.T. and Saw, J.G. (1972): Approximating the percentage points of Hotelling's generalized T_0^2 statistic. Biometrika, 59, 224-226.

χ^2 *and F approximations to Hotelling's* T_0^2 *distribution (null case). Percentage points. Tables.*

Hughes, H.M. (1960): See Danford, M.B., Hughes, H.M. and McNee, R.C., #211.

Hughes, J.B. (1962): See Smith, H., Gnanadesikan, R. and Hughes, J.B., #1045.

Hume, M.W. (1965): See Aitkin, M.A. and Hume, M.W., #14.

Hume, M.W. (1966): See Aitkin, M.A. and Hume, M.W., #16.

Hume, M.W. (1968): See Aitkin, M.A. and Hume, M.W., #17.

483. Huynh, H. and Feldt, L.S. (1970): Conditions under which mean square ratios in repeated measurements designs have exact F-distributions. J. Amer. Statist. Assoc., 65, 1582-1589.

Repeated measurements experiments: randomized block and split-plot designs. Character of covariance matrix which results in exact F-distributions for treatments and interaction variance ratios. Necessary and sufficient conditions are the equality of variances of differences for all pairs of treatment measures (assumed correlated). Equivalent to Box-Geisser-Greenhouse parameter $\varepsilon = 1.0$. Sufficient but not necessary condition: equality of variances and covariances. Sphericity test as a preliminary test on sufficient condition.

484. Ifram, A.F. (1970): On mixtures of distributions with applications to estimation. J. Amer. Statist. Assoc., 65, 749-754.

Mixtures of distributions. Mixture of functions of random variables. Noncentral χ^2 and F derived from normal distribution. Representations for total and multiple correlation coefficients. Estimation by principle of conditional moments. Application to estimation of noncentrality parameters.

485. Ihm, P. (1959): Numerical evaluation of certain multivariate normal integrals. Sankhya, 21, 363-366.

Evaluation of MVN integrals with covariance matrix $D + ii'/c^2$. MVN distribution with density constant over the surface of a hyperellipsoid, rotationally symmetric about longest axis.

486. Ihm, P. (1961): A further contribution to the numerical evaluation of certain multivariate normal integrals. Sankhya, A, 23, 205-206.

Extension of #485 to MV integral involving a complex argument.

487. Ishii, G. (1968): Minimax invariant prediction regions. Ann. Inst. Statist. Math., 20, 33-53.

Measure of desirability for region: reciprocal of volume (or area) of region. Risks: average probability content and average volume of region. Application to multivariate linear model. Minimax prediction region for n future observations after having observed m.

Isii, K. (1967): See Nakajima, N. and Isii, K., #830.

488. Ito, K. (1960): Asymptotic formulae for the distribution of Hotelling's generalized T_0^2 statistic II. Ann. Math. Statist., 31, 1148-1153.

Characteristic functin of non-null T_0^2 distribution. Expansion and inversion of c.f. to obtain asymptotic expansion for c.d.f. of T_0^2 in non-null case. Possible application in the evaluation of the power of test based on T_0^2.

489. Ito, K. (1961): On multivariate analysis of variance tests. Bull. Inst. Internat. Statist., 38, 87-98.

Comparison of various tests proposed for MANOVA: (i) Wilks' U and W, (ii) Lawley's V and Hotelling's T_0^2, (iii) Roy's largest root. Tests illustrated with example. Contradictions in results discussed.

490. Ito, K. (1962): A comparison of the powers of two multivariate analysis of variance tests. Biometrika, 49, 455-462. [Correction: 50 (1963), 546].

Comparison of powers of Hotelling's T_0^2 and Wilks' W. Test of $\mu_1 = \mu_2 = \ldots = \mu_k$ using MANOVA. Asymptotic results. Expansion of distribution functions of statistics. Table of power comparison.

491. Ito, K. and Schull, W.J. (1964): On the robustness of the T_0^2-test in multivariate analysis of variance when the variance-covariance matrices are not equal. Biometrika, 51, 71-82.

Asymptotic distribution of T_0^2 when assumption of equality of variance matrices not valid. Moments of the statistic. Approximation by central chi-square with c and f to be determined. Two moments method. Numerical results on α and power. Tables.

492. Ito, K. (1966): On the heteroscedasticity in the linear normal regression model. Res. Papers Statist. Festschr. Neyman, [David, ed.), 147-155.

MV linear model. A test for linearity of regression. MV extension of the UV test. Distribution of the criterion when homoscedasticity does not hold. Moments of the criterion. Approximation by χ^2. Determination of c and f by moments for large samples. Level of significance.

493. Ito, K. (1969): On the effect of heteroscedasticity and non-normality upon some multivariate test procedures. Multivariate Anal.-II, [Proc. 2nd Internat. Symp., Krishnaiah, ed.], 87-120.

Effects of heteroscedasticity and non-normality on the tests of hypotheses concerning the mean and the variance matrices in MV analysis. (i) Means: (a) extension and generalization of Scheffe-Bennett-Anderson class of procedures. Power comparisons. Optimal procedure when $\Sigma_1 = \theta\Sigma_2$. *Approximate procedures based on James' asymptotic expansion methods. Numerical evaluations and comparisons based on* α. *(b) Approximate methods in the non-normal case. Asymptotic approximations by chi-square with c and f to be determined by moments. Asymptotic moments of the statistics. (ii) Variance matrices: Tests of the hypotheses* $\Sigma = \Sigma_0$, $\Sigma = \sigma^2 I$, $\Sigma_{ij} = 0$, $\Sigma_1 = \Sigma_2 = \ldots = \Sigma_k$ *(k samples). LR tests. Effects of non-normality on the limiting distributions of the criteria. Asymptotic moments.*

494. Jack, H. (1965): Jacobians of transformations involving orthogonal matrices. Proc. Roy. Soc. Edinburgh, Sect. A, 67, 81-103.

Transformations involving orthogonal matrices. Jacobians. Volume of the set of all orthogonal matrices. Macbeath's technique. Matrix transformations and Wishart-Siegel theorems. Beta and Gamma functions of matrix argument. Related integrals.

495. Jackson, E.C. (1968): Missing values in linear multiple discriminant analysis. Biometrics, 24, 835-844.

An example of missing values in linear discriminant analysis. Estimation procedures using mean values or iterative regression technique. Application to data. Suggests equivalence of the procedures.

496. Jackson, J.E. and Morris, R.H. (1957): An application of multivariate quality control to photographic processing. J. Amer. Statist. Assoc., 52, 186-199.

Application of principal component analysis and its modifications. Interpretation of the components. Photographic data.

497. Jackson, J.E. (1959): Some multivariate statistical techniques used in color matching data. J. Opt. Soc. Amer., 49, 585-592. [Correction: 52 (1962), 835-836].

Use and interpretation of multivariate techniques for photographic data. Examples.

498. Jackson, J.E. (1959): Quality control methods for several related variables. Technometrics, 1, 359-377.

Control ellipses and ellipsoids. Use of T^2 on the original and on the principal variables. T^2 chart. Examination of "out-of-control" processes. Use and abuse of characteristic vector procedures for $p > 2$.

499. Jackson, J.E. and Bradley, R.A. (1961): Sequential χ^2 and T^2 tests. Ann. Math. Statist., 32, 1063-1077.

Sequential tests on the mean in MVN distribution, based on chi-square (Σ known) or T^2 (Σ unknown). SPRT. Cox's conditions and verification of their fulfillment in the MVN. Stein's technique and direct method. Termination proofs. Extensions to two sample tests. ASN. Application and extension of Bhate's method. Generalizations.

500. Jackson, J.E. and Bradley, R.A. (1961): Sequential χ^2 and T^2 tests and their application to an acceptance sampling problem. Technometrics, 3, 519-534.

Application of #499 to data. Some examples. Discussion of the uses of sequential procedures.

501. Jackson, J.E. and Bradley, R.A. (1966): Sequential multivariate procedures for means with quality control applications. Multivariate Anal.-I, [Proc. 1st Internat. Symp., Krishnaiah, ed.], 507-519.

Sequential procedures in #499 are further discussed. Some light is thrown on the alternative hypothesis and the use of principal component analysis prior to testing sequential. An example is given.

Jaiswal, M.C. (1963): See Khatri, C.G. and Jaiswal, M.C., #589.

502. Jaiswal, M.C. and Khatri, C.G. (1968): Estimation of paramters for the selected samples from bivariate normal populations. Metron, 27, 326-333.

Truncated BVN with truncation along X-axis. Moments method of estimation for mean and variance of X. ML estimation and relationship with moments method.

James, A.T. (1958): See Constantine, A.G. and James, A.T., #183.

503. James, A.T. (1960): The distribution of the latent roots of the covariance matrix. Ann. Math. Statist., 31, 151-158.

Distribution of the latent roots of a Wishart matrix with arbitrary covariance matrix. Evaluation of ∫ etr(BHAH')d(H) over the orthogonal group O(k). Zonal polynomials. Expansions. Distribution of the roots in series of zonal polynomials. Tables of zonal polynomials.

504. James, A.T. (1961): Zonal polynomials of the real positive definite symmetric matrices. Ann. Math. Ser. 2, 74, 456-469.

A discussion of zonal polynomial algebra and its relation to subalgebra of symmetric group. Isomorphisms.

505. James, A.T. (1961): The distribution of noncentral means with known covariance. Ann. Math. Statist., 32, 874-882.

Zonal polynomials and their calculation. Integrals involving zonal polynomials. Distribution of the roots of $|S - \lambda\Sigma| = 0$ in the noncentral case. Non-null Wishart distribution. Expansions in zonal polynomials.

506. James, A.T. (1964): Distribution of matrix variates and latent roots derived from normal samples. Ann. Math. Statist., 35, 475-497.

Functions of matrix argument: Hypergeometric and related polynomials involving matrix arguments. A compendium of MV analogues to UV distributions. The complex MVN. Zonal polynomials and their computation. Laguerre polynomials. Moments of Wishart matrix and other related statistics. Bibliographic.

507. James. A.T. (1966): Inference on latent roots by calculation of hypergeometric functions of matrix argument. Multivariate Anal.-I, [Proc. 1st Internat. Symp,, Krishnaiah, ed.], 209-235.

Likelihood function and its numerical evaluations. Principal components. Elimination of nuisance parameters by quasi-sufficiency and Bayesian (uniform prior) methods. Inferences based on likelihood. Expansion of likelihood in asymptotic and power series for three-variate case. Effects on LR of adjacent roots, of the distant roots. Example. Tables.

508. James, A.T. (1968): Calculation of zonal polynomials coefficients by use of the Laplace-Beltrami operator. Ann. Math. Statist., 39, 1711-1719.

Laplace-Beltrami operator and its eigenfunctions. Zonal polynomials and their calculation. Examples.

509. James, A.T. (1969): Tests of equality of latent roots of the covariance matrix. Multivariate Anal.-II, [Proc. 2nd Internat. Symp., Krishnaiah, ed.], 205-218.

Subasymptotic expansions for the joint distribution of the roots. Asymptotic expansion of tr(AHLH') and the integral of etr(-AHLH'). Linkage and non-independence of the roots. Conditional (asymptotic) distribution of the last (m-k) given the first k roots. Test for equality of the last (m-k) roots. Alternative proof of Lawley's result. The trivariate case.

510. James, W. and Stein, C.M. (1961): Estimation with quadratic loss. Proc. Fourth Berkeley Symp. Math. Statist. Prob., 1, 361-379.

MVN case with identity covariance matrix and quadratic loss. Computation of the best estimator when σ^2 is known and when the covariance matrix is unknown. Inadmissibility when $k \geq 3$. Estimation of Σ for $k \geq 2$ with known expectation. Best estimator invariant under full linear group but not minimax for a variety of loss functions. Minimax estimator for one choice of loss function. Estimator which is best invariant under the group of lower triangular matrices.

Jarnagin, M.P. (1961): See DiDonato, A.R. and Jarnagin, M.P., #259.

Jarnagin, M.P. (1962): See DiDonato, A.R. and Jarnagin, M.P., #260.

Jayachandran, K. (1967): See Pillai, K.C.S. and Jayachandran, K., #887.

Jayachandran, K. (1968): See Pillai, K.C.S. and Jayachandran, K., #890.

Jayachandran, K. (1970): See Pillai, K.C.S. and Jayachandran, K., #898.

511. Jeffers, J.N.R. (1962): Principal component analysis of designed experiment. Statistician, 12, 230-242.

Reduction of data by principal component analysis. Search for "dimensions" of variability. Application to designed experiment. Analysis of treatment means on the basis of principal components. Interpretation of principal components. Discussion.

512. Jeffers, J.N.R. (1967): Two case studies in the application of principal component analysis. Appl. Statist., 16, 225-236.

Uses of principal component analysis in the interpretation of data. Basic dimensions of variability. Choosing variables for further study. Applications to (i) problem in timber industry, (ii) classification of winged aphids.

513. Jenden, D.J., Fairchild, M.D., Mickey, M.R., Silverman, R.W. and Yale, C. (1972): A multivariate approach to the analysis of drug effects on the electroencephalogram. Biometrics, 28, 73-80.

Combination of multiple discriminant function analysis and canonical analysis. Drug effect on EEG. Continuous analyses of EEG over periods of several hours.

514. Jennings, E. (1965): Matrix formulas for part and partial correlations. Psychometrika, 30, 353-356.

General formulas for part and partial correlations of any order derived in terms of multiple correlation coefficients, standard partial regression weights and validities.

515. Jennrich, R.I. (1970): An asymptotic χ^2 test for the equality of two correlation matrices. J. Amer. Statist. Assoc., 65, 904-912.

Test of equality of two correlation matrices based on a function of the statistic R_1-R_2. Contradiction of Kullback's results. Adequacy of approximation. Simulation study. Application to study of factor patterns.

516. Jensen, D.R. (1969): An inequality for a class of bivariate chi-square distributions. J. Amer. Statist. Assoc., 64, 333-336.

Construction of two-sided simultaneous confidence limits for variance in the bivariate case. Develops inequality: $Pr\{c_1 << U_1 \le c_2, c_1 \le U_2 \le c_2\} \le Pr(c_1 \le U_1 \le c_2)Pr(c_1 \le U_2 \le c_2)$ when U_1 and U_2 have a bivariate chi-square distribution.

517. Jensen, D.R. and Jones, M.Q. (1969): Simultaneous confidence intervals for variances. J. Amer. Statist. Assoc., 64, 324-332.

Simultaneous confidence intervals for $\sigma_1^2, \sigma_2^2, \ldots, \sigma_p^2$ based on $U_i = \nu_i s_i^2/\sigma_i^2$ for $i = 1, \ldots, p$. Joint distribution of U_i's: multivariate χ^2 distribution. Bonferroni intervals. Lower bound based on inequality which relates *a joint distribution function to the product of its marginals. Comparison for $p = 2$ of Bonferroni-type intervals and those suggested by (i) Anderson and (ii) Roy and Gnandesikan.*

518. Jensen, D.R. (1970): The joint distribution of traces of Wishart matrices and some applications. Ann. Math. Statist., 41, 133-145. [Correction: 41 (1970), 2186].

Joint distribution of $(\nu_1, \ldots, \nu_q)$ where $\nu_j = tr(W_{jj}\Sigma_{jj}^{-1})$, W_{jj}, Σ_{jj} being diagonal blocks of partitioned Wishart W and its variance matrix Σ, respectively. Laguerre polynomials of vector argument. Multivariate χ^2 and F-distributions. Some inequalities on these distributions using Khatri's results. Applications to simultaneous inference. Lawley-Hotelling statistics.

519. Jensen, D.R. (1970): The joint distribution of quadratic forms and related distributions. Austral. J. Statist., 12, 13-22.

BV χ^2 and F distributions. Characteristic functions and their expansions in series involving Laguerre polynomials. Lancaster's canonical form.

520. Jensen, D.R. (1972): The limiting form of the noncentral Wishart distribution. Austral. J. Statist., 14, 10-16.

Limiting distribution of the standardized noncentral Wishart as the n.c. parameter θ tends to infinity. Characteristic function technique. MV chi-square and Rayleigh distributions.

521. Jensen, D.R. (1972): Some simultaneous multivariate procedures using Hotelling's T^2 statistics. Biometrics, 28, 39-53.

Joint distribution of Hotelling's T^2 statistics derived from partitioned vector. Inequalities for mean vector. Ellipsoidal regions in linear subspaces. Applications. Simultaneous inferences in the k sample case. Example. Analysis of repeated measurements. Axial symmetry in blocks. Test of hypothesis of the type H_0: $\mu_1 = \mu_2 = \ldots = \mu_t$ when all the μ's are vectors. Bonferroni type inequalities.

522. Jogdeo, K. (1970): A simple proof of an inequality for multivariate normal probabilities of rectangles. Ann. Math. Statist., 41, 1357-1359.

A simple proof of Sidak's result on the probability of rectangles in MVN. Directional derivatives.

523. John, J.A. (1970): Use of generalized inverse matrices in MANOVA. J. Roy. Statist. Soc., B, 32, 137-143.

Generalization of Rao's and Searle's results to MANOVA. Application of generalized inverse. Estimable functions, MLE, LR test, general linear hypothesis, one and two way classifications.

524. John, S. (1959): On the evaluation of the probability integral of a multivariate normal distribution. Sankhya, 21, 367-370.

Expressing MVN probability for n dimensions in terms of integrals involving (n-1) dimensions. General case. Applications.

525. John, S. (1959): The distribution of Wald's classification statistic when the dispersion matrix is known. Sankhya, 21, 371-376.

Distribution of $W = (\bar{X}_1-\bar{X}_2)'\ ^{-1}y$ *where* $(\bar{X}_1-\bar{X}_2)$ *is MVN independent of Y and* Σ, *the common variance matrix, is known. Characteristic function and its expansion. Inversion of the characteristic function.*

526. John, S. (1960): On some classification problems-I. Sankhya, 22, 301-308. [Correction: 23A (1961), 308].

A new classification procedure. Probability of mis-classification. Bounds on the probability. Extensions to more than two populations.

527. John, S. (1960): On some classification statistics. Sankhya, 22, 309-316.

The exact distribution of the classificatory statistics of Wald, Anderson and Rao in the general case when the variance matrix is assumed equal and known.

528. John, S. (1961): Errors in discrimination. Ann. Math. Statist., 32, 1125-1144.

Probability of misclassification is discussed. Its distribution is obtained for (i) only μ_2 *unknown, (ii) only* Σ *known, (iii)* μ_1, μ_2, Σ *all unknown. Expected value of the probability of misclassification. Asymptotic distributions.*

529. John, S. (1961): On the evaluation of the probability integral of the multivariate t-distribution. Biometrika, 48, 409-417.

MV t-distribution with correlation matrix P. Characteristic function. Probability integral for p = 2. Infinite series expansion in powers of correlation coefficient ρ and integral involving Hermite polynomials. Recurrence relations. MV generalization. Applications: SCB's (i) on mean vector of MVN when variances are equal or ratios of variances are known, (ii) for a future observation, (iii) on regression coefficients in univariate case. Tables.

530. John, S. (1962): On classification by the statistics R and Z. Ann. Inst. Statist. Math., 14, 237-246. [Correction: 17 (1965), 113].

Probabilities of misclassification when the classification is by Rao's R or John's Z statistic. Unknown variance matrix case. Distribution and expected values of the probabilities.

531. John, S. (1963): A tolerance region for multivariate normal distributions Sankhya, A, 25, 363-368.

Tolerance regions for MVN. Large sample result. Procedures based on geometric mean and the harmonic mean. Bounds on the probability. Σ unknown but of the form $\sigma^2 \Lambda$ with Λ known.

532. John, S. (1964): Further results on classification by W. Sankhya, A, 26, 39-46.

Classification of an individual from P into one or other of P_1, P_2. Distribution and expectation of error of classification when (i) only μ_2 unknown, (ii) both μ_1, μ_2 unknown. Large sample result for all parameters unknown.

533. John, S. (1964): Methods for the evaluation of probabilities of polygonal and angular regions when the distribution is bivariate t. Sankhya, A, 26, 47-54.

Integral form for the probability. Formulae for the integral (i) in terms of infinite series involving incomplete Beta, (ii) in powers of 1/n, (iii) in a closed form, (iv) in an approximate form. A recurrence relation. Application to testing for outliers.

534. John, S. (1966): On the evaluation of probabilities of convex polyhedra under multivariate normal and t distributions. J. Roy. Statist. Soc., B, 28, 366-369.

Generalization of #533 to multidimensinnal case. Recurrence relations for $V(h_1,\ldots,h_p)$. Series expanxion in terms of $1/n$. MV t-distribution.

535. John, S. (1968): A central tolerance region for the multivariate normal distribution. J. Roy. Statist. Soc., B, 30, 599-601.

Three special cases looked at: Σ known, μ known, neither known. Simultaneous central tolerance intervals.

536. John, S. (1971): Some optimal multivariate tests. Biometrika, 58, 123-127.

Tests optimal for detecting small deviations from H_0. Locally best tests. Directional derivations of the power. Tests of one or two sided alternatives: (i) $\Sigma = I$, (ii) $\Sigma = \sigma^2\Sigma_0$, Σ_0 known, (iii) independence of sets of variates, (iv) $\Sigma_1 = \Sigma_2$, (v) linear hypothesis. Null distributions of the criteria considered. All tests based on directional differentials.

537. John, S. (1971): A test of equality of block-diagonal covariance matrices and its role of unification. J. Roy. Statist. Soc., B, 33, 301-306.

A general formulation of hypothesis testing for covariance matrices. LR criterion. Asymptotic distribution using Box's techniques. Central and noncentral χ^2 approximations. Generalizes several known results. Mellin transform techniques.

538. John, S. (1972): The distribution of a statistic used for testing sphericity of normal distribution. Biometrika, 59, 169-173.

Joint distribution of sum of the powers and the product of roots of a Wishart matrix. Null distribution of LR and John's (#537) criteria for $\Sigma = \sigma^2\Sigma_0$, Σ_0 known, σ^2 unknown. Approximations.

Johnson, R.A. (1968): See Bhattacharyya, G.K. and Johnson, R.A., #105.

Johnson, R.A. (1971): See Bhattacharyya, G.K., Johnson, R.A. and Neave, H.R., #106.

539. Jolicoeur, P. and Mosimann, J.E. (1960): Size and shape variation in the painted turtle: A principal component analysis. Growth, 24, 339-354.

Interpretation of principal components. Concepts of size and shape. Geometric interpretation of bivariate and trivariate principal components. Example: length, width, and height of male and female painted turtles.

540. Jolicoeur, P. (1963): The multivariate generalization of the allometry equation. Biometrics, 19, 497-499.

Multivariate extension of allometry equation: $y = bx^a$. Generalization: first principal component of the covariance matrix of logarithms. Isometry hypothesis. Positive and negative allometry. Test of hypothesis concerning first principal component.

541. Jolicoeur, P. (1968): Interval estimation of the slope of the major axis of a bivariate normal distribution in the case of a small sample. Biometrics, 24, 679-682.

Confidence interval for slope of major axis of BVN. Small sample procedure based on distribution of squared correlation coefficient. Interval covers the slopes of all hypothetical major axes such that each is not significantly correlated with corresponding minor axis. Comparison with large sample results. Examples.

542. Jolliffe, I.T. (1972): Discarding variables in a principal component analysis, I: Artificial data. Appl. Statist., 21, 160-171.

Rejecting "redundant" variables before undertaking a principal component analysis. Rejection techniques based on (i) multiple correlation, (ii) principal components, (iii) cluster analysis. Comparisons using artifical data and some examples from literature.

Jones, M.Q. (1969): See Jensen, D.R. and Jones, M.Q., #517.

543. Joshi, V.M. (1967): Inadmissibility of the usual confidence sets for the mean of a multivariate normal population. Ann. Math. Statist., 38, 1868-1875.

Confidence sets for estimating an unknown mean vector μ, $p \geq 3$, when X is MVN with known $\Sigma = I$. Stein's conjecture. Inadmissibility of usual confidence sets which are spheres of fixed volume centered at the sample mean.

544. Joshi, V.M. (1969): Admissibility of the usual confidence sets for the mean of a univariate or bivariate normal population. Ann. Math. Statist., 40, 1042-1067.

Continuation of #543. Stein's conjecture. Admissibility of confidence sets for $p = 1$ or $p = 2$. Equivalence classes. Confidence procedures with open or convex sets. Parallels Stein's results regarding point estimation of mean.

Jouris, G.M. (1969): See Pillai, K.C.S. and Jouris, G.M., #893.

Jouris, G.M. (1969): See Pillai, K.C.S., Al-Ani, S. and Jouris, G.M., #895.

Jouris, G.M. (1971): See Pillai, K.C.S. and Jouris, G.M., #900.

545. Jowett, G.H. (1963): Applications of Jordan's procedure for matrix inversion in multiple regression and multivariate distance analysis. J. Roy. Statist. Soc., B, 25, 352-357.

Partitioning Mahalanobis' D^2 statistic. Covariance matrix known. Conditional contribution to D^2. Use of Jordan's method for matrix inversion. Similarity with pivotal condensation method.

546. Juritz, J.M. (1971): The generalised partial correlation matrix. South African Statist. J., 5, 1-4.

Definition of generalised partial correlation matrix P. Other measures of partial correlation defined as special cases. MLE of P under assumption of MVN. Sampling distribution and moments of MLE. Testing that two sets of variates are conditionally independent assuming normality.

547. Kabe, D.G. (1958): Some applications of Meijer-G functions to distribution problems in statistics. Biometrika, 45, 578-580.

Meijer-G function and its relation to Nair's differential equation. Application to generalized correlation ratio and the joint distribution of the canonical correlations of (X_1, X_2) with $(Y_1, \ldots, Y_p)$.

548. Kabe, D.G. (1962): On the exact distribution of a class of multivariate test criteria. Ann. Math. Statist., 33, 1197-1200.

Distribution of a linear function of gamma variates. Characteristic function and its inversion. LR criterion for MANOVA. Characteristic function of $-2 \ln \Lambda$ and its frequency function for special values of parameters.

549. Kabe, D.G. (1963): Estimation of a set of fixed variates for observed values of dependent variates with normal multivariate regression models subjected to linear restrictions. Ann. Inst. Statist. Math., 15, 51-59.

Estimation of independent variates from an estimated MV regression model. Linear restrictions. Conditional distribution. Hotelling's T^2. Application of Sverdrup's lemma and its generalizations.

550. Kabe, D.G. (1963): Some results on the distribution of two random matrices used in classification procedures. Ann. Math. Statist., 34, 181-185. [Correction: 35 (1964), 924].

Generalization of Anderson-Sitgreaves result on the distribution of some classificatory statistics. Removal of the restriction of the proportionality of the mean vectors. Some integration results. Distributional properties of noncentral Wishart in general and for the planar case.

551. Kabe, D.G. (1963): Stepwise multivariate linear regression. J. Amer. Statist. Assoc., 58, 770-773.

MV linear model $Y = BX + E \equiv B_1X_1 + B_2X_2 + E$. Estimation of B in two steps as $(\hat{B}_1, \hat{B}_2)$ or as a composite matrix. Biases in the procedures. Test criterion for H_0: $B_2 = 0$. Distribution of the criterion. MV analog of Vail-Ross-Jochem-Goldberger results.

552. Kabe, D.G. (1963): Multivariate linear hypothesis with linear restrictions. J. Roy. Statist. Soc., B, 25, 348-351.

MV linear model $Y = BX + E$. MLE of B and Σ. LR test for the restrictions $FB' = W'$. The hth moment of the criterion. Estimation of the linear functions $MB' = Z$ under the restrictions $FB' = W'$. Conditional density of $\hat{Z}$ given $\hat{W}$. Sverdrup's lemma, its generalizations and applications.

553. Kabe, D.G. (1964): Decomposition of Wishart distribution. Biometrika, 51, 267.

An alternative proof for Teichroew-Sitgreaves' decomposition of Wishart (central) into normal and t variates. Direct transformation method.

554. Kabe, D.G. (1964): A note on the Bartlett decomposition of a Wishart matrix. J. Roy. Statist. Soc., B, 26, 270-273.

Bartlett decomposition established by factorization of the generalized variance. Direct transformation of variates. Simpler proof. Noncentral Wishart. Decomposition of p-dimensional of rank t into t dimensional of rank t, (p-t) independent chi-squares, and unit normals.

555. Kabe, D.G. (1964): On tests concerning coefficient matrices of linear restriction with normal regression models. Metrika, 8, 231-234.

LR criterion for testing the hypothesis H_0: $PB_1' = PB_2'$ when $Y_i = B_iX_i + E_i$, $i = 1, 2$. Moments of the LR statistic and its approximate distribution.

556. Kabe, D.G. (1965): On the noncentral distribution of Rao's U statistic. Ann. Inst. Statist. Math., 17, 75-80.

Alternative derivation of distribution of Rao's U statistic used for discriminant analysis. Non-null case. Use of Sverdrup's lemma and its generalizations.

557. Kabe, D.G. (1965): Generalization of Sverdrup's lemma and its applications to multivariate distribution theory. Ann. Math. Statist., 36, 671-676.

Sverdrup's lemma on integration with restrictions. Generalizations. Applications to matrix of regression coefficients, a classificatory statistic, and Hotelling's T^2.

Kabe, D.G. (1965): See Hayakawa, T. and Kabe, D.G., #451.

558. Kabe, D.G. (1966): Some results for the normal multivariate regression models. Austral. J. Statist., 8 22-27.

Sverdrup's lemma, its generalizations and uses. Combination of estimates for B in the models $Y_i = X_iB + E_i$, $i = 1, 2$. Joint distribution of $\hat{B}$ and G. Rejection criterion for outliers in MVN case.

559. Kabe, D.G. (1966): Complex analogues of some results in classical normal multivariate regression theory. Austral. J. Statist., 8, 87-91.

MV regression model in the complex case. Prediction of the dependent variable. Distribution of its error matrix under a restriction. Case of correlatedness of predicted and observed values. Estimation of X. Extension of Kabe's results.

560. Kabe, D.G. (1966): Complex analogues of some classical noncentral multivariate distributions. Austral. J. Statist., 8, 99-103.

Alternative derivations of the distributions of T^2, R^2, r and partial correlation coefficient in the complex case. Finite series representation of the densities. Direct integration method.

561. Kabe, D.G. (1966): On the distribution of the complex analog of Rao's U statistic. J. Indian Statist. Assoc., 4, 189-194.

Extension of Kabe's proof on Rao's U statistic to complex case. Complex analog of Hotelling's T^2. Application of U statistic for testing a regression hypothesis.

562. Kabe, D.G. (1967): On multivariate prediction intervals for sample mean and covariance based on partial observations. J. Amer. Statist. Assoc., 62, 634-637.

Prediction region for $\bar{Y}$ based on first k observations and T^2. Regions for $|S|$ based on the LR statistic U and $|S_1|$. Prediction of regression matrix.

563. Kabe, D.G. (1967): Minima of vector quadratic forms with applications to statistics. Metrika, 12, 155-160.

The generalization of Bush-Olkin results on the minima of quadratic forms to expressions of the type $tr[YHY']$ subject to $DY' = V'$ where Y is now a matrix. Several applications of the result to multivariate generalizations of known univariate results.

564. Kabe, D.G. (1968): On the distributions of direction and collinearity factors in discriminant analysis. Ann. Math. Statist., 39, 855-858.

Alternative derivations of the distributions of the factors Λ_1, Λ_2, Λ_3, Λ_4, Λ_5 in discriminant analysis. Direct evaluations without use of transformations defined by Kshirasagar.

565. Kabe, D.G. (1968): On the distribution of the regression coefficient matrix of a normal distribution. Austral. J. Statist., 10, 21-23.

Regression of X_2 on X_1 when MVN. Distribution of MLE of the regression matrix and the intercept vector. Marginal density of the regression matrix when $E(X) \neq 0$. Tests of hypotheses on the intercept being zero and on the regression matrix being zero. Test of H_0: $FB = W$.

566. Kabe, D.G. (1968): Some aspects of analysis of variance and covariance theory for a certain multivariate complex Gaussian distribution. Metrika, 13, 86-97.

MANOVA and MANCOVAR. Design of experiments with two way classification. Complex normal and T^2 distributions. Tests for treatment differences.

567. Kabe, D.G. (1968): Minimum variance unbiased estimate of a covarage probability. Operations Res., 16, 1016-1020.

MVUE of the coverage probability $Pr\{X \leq a\} = F(a, \theta)$ is derived when θ admits a complete sufficient statistic. Integral expressions for the coverage probabilities for MVN with (i) Σ unknown, (ii) μ unknown, (iii) μ, Σ unknown. No explicit evaluation in general for the integrals.

568. Kabe, D.G. (1969): Linear compounds of normal linear regression coefficients. J. Roy. Statist. Soc., B, 31, 426-431.

Sverdrup's lemma and its generalization. MV normal regression theory. Joint distributions of statistics associated with the regression analysis. Distribution of double linear compound (studentized). Confidence bounds based on the t distribution are narrower than those on Roy's largest root test. Linear compounds of the mean vector. F distribution of the standardized form. Bounds on linear compounds are narrower than those based on T^2.

Kabe, D.G. (1970): See Gupta, R.P. and Kabe, D.G., #417.

Kabe, D.G. (1971): See Gupta, R.P. and Kabe, D.G., #418.

Kabe, D.G. (1971): See Wani, J.K. and Kabe, D.G., #1154.

569. Kabe, D.G. (1972): On the noncentral distribution of a certain correlation matrix. South African Statist. J., 6, 27-29.

Non-null distribution of $Q(A+B)^{-\frac{1}{2}}B(A+B)^{-\frac{1}{2}}Q'$ where Q is an orthogonal matrix, A and B are Wishart. Density represented in a form involving hypergeometric function of matrix argument. Difficulty in finding the distribution of the matrix $(A+B)^{-\frac{1}{2}}B(A+B)^{-\frac{1}{2}}$.

570. Kabe, D.G. and Gupta, R.P. (1972): Decomposition of the multivariate beta distribution with applications. Canad. Math. Bull., 15, 225-228.

Decomposition of MV Beta (non-null) into independent Betas. Radcliffe-Pillai-Khatri decomposition. Application to discriminant analysis and the distribution of a factor. Noncental Wishart.

571. Kamat, A.R. (1958): Hypergeometric expansions for incomplete moments of the bivariate normal distribution. Sankhya, 20, 317-320.

Determination of incomplete moments [m,n] for BVN. Expectation over the positive quadrant. Derivation of complete absolute moments (m,n) in terms of [m,n]. Expansion of [m,n] in terms of a hypergeometric series in ρ^2. *Tables of [m,n] for* $m + n \leq 4$, $\rho = 0.1(0.1)1.0$ *and* $\rho = 0.0(-0.1)-0.9$.

572. Kamat, A.R. (1958): Incomplete moments of the trivariate normal distribution. Sankhya, 20, 321-322.

Extension of #571 to trivariate case. Expressions for incomplete moments [ℓ, m, n] for $\ell + m + n \leq 4$ *in terms of correlation coefficients.*

Kanazawa, M. (1968): See Okamoto, M. and Kanazawa, M., #847.

Kannel, W. (1967): See Truett, J., Cornfield, J. and Kannel, W., #1140.

573. Kappenman, R.F., Geisser, S. and Antle, C.E. (1970): Bayesian and fiducial solutions for the Fieller-Creasy problem. Sankhya, B, 32, 331-340.

Interval estimation for $\eta = \mu_2/\mu_1$ *in BVN problem: (i) Bayesian solution using non-informative prior density for* μ *and* Σ^{-1}, *(ii) extension of Creasy's solution using conditional distribution for* μ. *Computer programs for determining posterior densities and intervals. Example.*

574. Kappenman, R.F. (1971): A note on the multivariate t-ratio distribution. Ann. Math. Statist., 42, 349-351.

Distribution of $(x_2/x_1, x_3/x_1, \ldots, x_p/x_1)$ *when* $x_1, x_2, \ldots, x_p$ *has a MV t-distribution. Density when p is even and odd. Multivariate generalization of Fieller-Creasy problem.*

Kappenman, R.F. (1971): See Geisser, S. and Kappenman, R.F., #347.

575. Karlin, S. and Truax, D. (1966): Slippage problems. Ann. Math. Statist., 31, 296-324.

Slippage in location parameter in a MVN. Symmetric procedures based on X and S. Invariant under nonsingular transformations. Monotone likelihood ratio. Distribution of maximal invariant. Bayes procedures and their characterizations.

576. Katti, S.K. (1961): Distribution of the likelihood ratio for testing multivariate linear hypotheses. Ann. Math. Statist., 32, 333-335.

Distribution of LR for testing multivariate linear hypotheses: product of q independent Beta variables. Random orthogonal transformations.

Kaufman, G.M. (1965): See Ando, A. and Kaufman, G.M., #45.

577. Kempthorne, O. (1966): Multivariate responses in comparative experiments. Multivariate Anal.-I, [Proc. 1st Internat. Symp., Krishnaiah, ed.], 521-539.

Review article on the pitfalls and drawbacks of multivariate analysis.

Kendall, M.G. (1958): See Daniels, H.E. and Kendall, M.G., #212.

578. Kendall, M.G. (1960): The evergreen correlation coefficient. Contrib. Prob. Statist., (Hotelling Volume, Olkin et al, eds.), 274-277.

Review of distribution and moments of simple correlation coefficient. Geometric interpretations. Impact of Hotelling's work on correlation coefficient: theory of transformations and theory of unbiased estimating functions.

579. Kendall, M.G. (1966): Discrimination and classification. Multivariate Anal.-I, [Proc. 1st Internat. Symp., Krishnaiah, ed.], 165-185.

Distinction between classification and discrimination. Nonparametric techniques applicable to data which are measurable, categorized or a mixture of both. Convex-hull method. Order statistic method. Examples.

Kendall, M.G. (1967): See Beale, E.M.L., Kendall, M.G. and Mann, D.W., #82.

Kerridge, D.F. (1967): See Day, N.E. and Kerridge, D.F., #236.

580. Kettenring, J.R. (1971): Canonical analysis of several sets of variables. Biometrika, 58, 433-451.

Extension of canonical correlation analysis to more than two variables. Isolation and summarization of linear relations among sets of variables. Optimization of some function of correlation matrix. Several methods considered: (i) maximum variance, (ii) minimum variance, (iii) sum of correlations, (iv) sum of squared correlations, (v) generalized variance method. Iterative procedures for generating canonical variates using methods (iii), (iv) and (v). Example.

Kettenring, J.R. (1972): See Gnanadesikan, R. and Kettenring, J.R., #384.

581. Khakhubiya, Ts.G. (1965): A lemma on random determinants and its application to the characterization of multivariate distributions. Theory Prob. Appl., 10, 685-689.

Distribution of $D = \det\{v_{ij}\}$, where v_{ij} are independently distributed random variables. Characterization of MVN: sufficient conditions which uniquely establish distributions of v_{ij} from distribution of D.

582. Khakhubiya, Ts.G. (1966): On random determinants. Theory Prob. Appl., 11, 185-186:

Review of literature on random determinants.

583. Khan, R.A. (1968): Sequential estimation of the mean vector of a multivariate normal distribution. Sankhya, A, 30, 331-334.

Sequential estimation of μ when Σ is a diagonal matrix of unknown elements. Stopping rule under quadratic loss function. Sequential confidence region for μ, which will ensure bounds on lengths of axes of the ellipsoid with given coverage probability.

584. Khatri, C.G. (1959): On the mutual independence of certain statistics. Ann. Math. Statist., 30, 1258-1262. [Correction: 32 (1961), 1344].

Wishart distribution. Transformations and Jacobians. Independence of $S_1 = S + XX'$ and $Z = T^{-1}X$ where $S = TT'$. Joint density of S_1 and Z. Mutual independence of characteristic roots. Application to stepwise discriminant analysis (Rao's U statistic). Simultaneous tests and simultaneous confidence intervals.

585. Khatri, C.G. (1959): On conditions for the forms of the type XAX' to be distributed independently or to obey Wishart distribution. Calcutta Statist. Assoc. Bull., 8, 162-168.

Generalization of results on quadratic forms to MV case: XAX', where X is a pxn matrix. Necessary and sufficient conditions for the Wishartness of XAX'. Determination of m.g.f. and c.g.f. of XAX'. Conditions for independence of XAX' and XBX' and of XAX' and XL. Idempotent matrices.

586. Khatri, C.G. (1961): A note on the interval estimation related to the regression matrix. Ann. Inst. Statist. Math., 13, 145-146.

Simultaneous confidence bounds for regression matrix $\beta = \Sigma_{12}\Sigma_{22}^{-1}$ *for the q-set fixed and stochastic. Demonstration that Roy's procedure yields narrower bounds than that of Siotani.*

587. Khatri, C.G. (1961): Simultaneous confidence bounds on the departures from a particular kind of multicollinearity. Ann. Inst. Statist. Math., 13, 239-242.

Departures from multicollinearity of regression coefficients. Construction of simultaneous confidence bounds on $\delta_{1.2} = \beta_{1.2} - \beta_{1.2}^0$ *where* $\beta_{1.2} = \beta_1 - \Sigma_{12}\Sigma_{22}^{-1}\beta_2$ *and* $\beta_{1.2}^0$ *is a given matrix. Distribution of maximum characteristic root in null case.*

588. Khatri, C.G. (1962): Conditions for Wishartness and independence of second degree polynomials in a normal vector. Ann. Math. Statist., 33, 1002-1007.

Considers matrix whose elements are second degree polynomials in a normal vector: $XAS' + \frac{1}{2}(LX' + XL') + C$. *Necessary and sufficient conditions for Wishartness and independence of such matrices. Extension of results given for XAX' in #585.*

589. Khatri, C.G. and Jaiswal, M.C. (1963): Estimation of parameters of a truncated bivariate normal distribution. J. Amer. Statist. Assoc., 58, 519-526.

Estimates of parameters of a singly truncated BVN using eight (or nine) moments. Asymptotic variances and covariances of estimates. ML estimates. Use of MM estimates as starting values of iterative solution for ML estimates. Asymptotic efficiency based on generalized variance. Example.

590. Khatri, C.G. (1963): Further contributions to Wishartness and independence of second degree polynomials in normal vectors. J. Indian Statist. Assoc., 1, 61-70.

NSC on the Wishartness and independence of second degree polynomials of the type $XAX' + (LX' + XL')/2 + C$, *where* X *is a matrix of MVN vectors. Noncentral, singular cases dealt with. Independence of (i) two polynomials (ii) linear and second degree polynomial. Applications to independence of several statistics including* $\bar{X}$ *and* $\sup[(x_i-\bar{x})A(x_j-\bar{x})']$ *over* (i,j). *Generalization of the results (for univariate case) of Bhat.*

591. Khatri, C.G. (1963): Joint estimation of the parameters of multivariate normal populations. J. Indian Statist. Assoc., 1, 125-133.

Estimation of the common mean of several MVN populations having different variance matrices. ML and Kullback's minimum discrimination information. Iterative solutions. Comparison of results. Numerical examples.

592. Khatri, C.G. (1964): Distribution of the 'generalized' multiple correlation matrix in the dual case. Ann. Math. Statist., 35, 1801-1806.

MV beta and pseudo MV beta distributions in the non-null case. Generalizations of multiple correlation to matrix form. Duality of relationship between the distributions of multiple correlation matrix. Null and non-null distributions of this matrix. Application to discriminant analysis. Tests on direction and collinearity factors of a hypothetical discriminant function in dummy variables.

593. Khatri, C.G. (1964): Distribution of the largest or the smallest characteristic root under null hypothesis concerning complex multivariate normal populations. Ann. Math. Statist., 35, 1807-1810.

The distribution function of the largest (or smallest) root of a matrix. Null distribution obtained by integration and expansion of Vander Monde's determinant. Incomplete beta and gamma integrals.

594. Khatri, C.G. (1965): A note on the confidence bounds for the characteristic roots of dispersion matrices of normal variates. Ann. Inst. Statist. Math., 17, 175-183.

Simultaneous confidence bounds on characteristic roots: (i) max ch Σ, min ch Σ, (ii) max ch $\Sigma_1\Sigma_2^{-1}$, min ch $\Sigma_1\Sigma_2^{-1}$, (iii) sth root of Σ_i, $i = 1,2,\ldots,k$, $s = 1, p$. Comparison with Anderson, Gnanadesikan and Roy procedures. Derivations based on Bartlett decomposition. χ^2 and F distributions. Union-intersection principle.

595. Khatri, C.G. (1965): Classical statistical analysis based on a certain multivariate complex Gaussian distribution. Ann. Math. Statist., 36, 98-114.

Jacobians of some transformations in complex case. MLE of mean and variance matrix. Their joint distribution and independence. LR criterion. MANOVA and other criteria. Partial vectors. Noncentral MV beta in complex case. Noncentral complex Wishart. Generalized variance, its moments. Distribution of Wishart in the linear case. Asymptotic distribution under the hypothesis. Regression. Characteristic roots of Hermitian matrices. Several cases considered.

596. Khatri, C.G. (1965): A test for reality of a covariance matrix in a certain complex Gaussian distribution. Ann. Math. Statist., 39, 115-119.

Tests on covariance structure in complex MVN. Test for $\Sigma_2 = 0$ when $\Sigma = \Sigma_1 + i\Sigma_2$. Test criteria: (i) LR criterion, (ii) $tr[S_2S_1^{-1}S_2'(S_1-S_2S_1^{-1}S_2')^{-1}]$, (iii) maximum characteristic root of $[S_2S_1^{-1}S_2'(S_1-S_2S_1^{-1}S_2')^{-1}]$. Distribution of criteria under H_0.

597. Khatri, C.G. (1965): Non-null distribution of the likelihood ratio statistic for a particular kind of multicollinearity. J. Indian Statist. Assoc., 3, 212-219.

Multicollinearity of regression coefficients. Test of H_0: $\beta_{1.2}Z = \beta^0_{1.2}Z$ (given). Non-null distribution of LR statistic. Reduction of certain problem to the stated H_0: (i) testing linear hypothesis for the departure from multicollinearity of regression coefficients, (ii) testing equality of the departure from multicollinearity of regression matrices with (a) $\beta_{1.2}$ unknown, (b) $\beta_{1.2}$ known and equal to $\beta^0_{1.2}$.

598. Khatri, C.G. and Pillai, K.C.S. (1965): Some results on the noncentral multivariate beta distribution and moments of traces of two matrices. Ann. Math. Statist., 36, 1511-1520.

Distribution of the matrix L with $A_2 = CYY'C' = CLC'$ such that $A_1 + A_2 = CC'$; A_1 has Wishart distribution and A_2 has a (pseudo) noncentral (linear) Wishart distribution. Distribution of L is expressible as the product of independent betas: the multivariate beta distribution. First and second moments of $U^{(s)}$ (Hotelling's generalized T_0^2) and Pillai's $V^{(s)}$ in the noncentral linear case.

599. Khatri C.G. (1966): A note on a MANOVA model applied to problems in growth curve. Ann. Inst. Statist. Math., 18, 75-86.

Application of generalized MANOVA to growth curve models (Potthoff and Roy). LR test of $C\eta V = 0$ or equivalently $C\xi V = 0$. Heuristic procedures for testing based on roots of matrix occurring in LR criterion. Non-null distribution of criteria. Monotonicity of the power. Simultaneous confidence bounds for $C\eta V$. Numerical example based on data from Potthoff and Roy.

600. Khatri, C.G. (1966): A note on a large sample distribution of a transformed multiple correlation coefficient. Ann. Inst. Statist. Math., 18, 375-380.

Transformed multiple correlation coefficient $u = gR^2/(1-R^2)$ and $g = g(\rho^2, n, k)$ under assumption (i) u is approximately noncentral F with 2k and 2n d.f. and noncentrality parameter δ or, (ii) u is approximately central F with ν and 2n d.f. where ν is a function of ρ^2, n, k. Obtain (g, δ) and (g, ν) by equating first two moments of u with those of approximating distribution. Tables showing accuracy of approximations for various n, k, ρ^2.

601. Khatri, C.G. (1966): On certain distribution problems based on positive definite quadratic functions in normal vectors. Ann. Math. Statist., 37, 468-479.

Zonal polynomials and integrals involving zonal polynomials. Distribution of (i) XLX', (ii) $Y'(XLX')^{-1}Y$, (iii) characteristic roots of $(YY')(XLX')^{-1}$. Moments of certain statistics. Similar results for complex MVN.

602. Khatri, C.G. and Pillai, K.C.S. (1966): On the moments of the trace of a matrix and approximations to its noncentral distribution. Ann. Math. Statist., 37, 1312-1318.

Extend results of #598. Obtain third and fourth moments of Hotelling's $U^{(p)}$. *Approximations for noncentral (linear) distribution of* $U^{(p)}$ *based on equating moments. Pearson type distribution. Comparison of approximation with exact distributions for* $p = 1$ *and* 2.

603. Khatri, C.G. (1967): Some distribution problems connected with the characteristic roots of $S_1S_2^{-1}$. Ann. Math. Statist., 38, 944-948.

Expression for $\int|I+\Lambda^{-1}HFH'|^{-t}dH$, *integrated over an orthogonal group (normalized to unity). Zonal polynomials and hypergeometric functions of matrix argument. Joint distributions of the roots* $(f_1,\ldots,f_p)$ *of* $S_1S_2^{-1}$ *and of the roots of* $S_1(S_1+S_2)^{-1}$. *Trace* T *of* $S_1S_2^{-1}$ *and its moment generating function. Distribution of* T. *Maximum root* f_p. *The joint distribution of the ratios* $x_i = f_i/f_p$, $i = 1,\ldots,p-1$. *Trace of* $S_1(S_1+S_2)^{-1}$. *Null distribution of* $(y_1,\ldots,y_{p-1})$, *where* $y_i = (f_i-f_1)/(f_p-f_1)$, $i = 1, 2,\ldots,p-1$. *Tests of hypothesis* H_0: $\Sigma_1\Sigma_2^{-1} = \lambda I$.

604. Khatri, C.G. (1967): On certain inequalities for normal distributions and their applications to simultaneous confidence bounds. Ann. Math. Statist., 38, 1853-1867.

Symmetric and separate symmetric regions. Probability contents of convex symmetric regions. $P\{\Pi X_j\} \geq \Pi P\{X_j\}$ *for a special structure of variance matrix. SCB on variances in MVN and on the ratios of variance in the two sample case. Maximum or minimum of several* T^2 *statistics. Probability bounds. SCB on locating parameters for special covariance structure. MANOVA growth curve model. Monotonicity of some probabilities involving MVN and Wishart distributions.*

605. Khatri, C.G. (1967): A theorem on least squares in multivariate linear regression. J. Amer. Statist. Assoc., 62, 1494-1495.

Optimality of LSE of the regression coefficient matrix: (i) maximizes the sample canonical correlations between Y *and linear transformations of* X, *(ii) maximizes all the characteristic roots of the regression sum of squares and products matrix for regression of* Y *on linear transformations of* X. *Generalization of Chow's results.*

606. Khatri, C.G. and Pillai, K.C.S. (1967): On the moments of traces of two matrices in multivariate analysis. Ann. Inst. Statist. Math., 19, 143-156.

Moments of $V^{(p)}$ and $U^{(p)}$. Evaluation of the moments of $V^{(p)}$ in terms of those of $V^{(r)}$, r being number of non-null roots of $\Sigma^{-1}\mu\mu'$. Zonal polynomials. $V^{(2)}$ and its moments.

607. Khatri, C.G. (1968): A note on exact moments of arcsine correlation coefficient with the help of characteristic function. Ann. Inst. Statist. Math., 20, 143-149.

Differential difference equation for the characteristic function of arcsine r. Expressions for the characteristic function. Exact moments. Moment relations and their solutions.

608. Khatri, C.G. (1968): Some results for the singular normal multivariate regression models. Sankhya, A, 30, 267-280.

Conditional of Y given X singular (due to Σ being singular). MLE of the regression matrix and Σ. Tests of hypotheses concerning the regression matrix: (i) H_0: $C\beta L = 0$, (ii) H_0: $E(Y) = \beta X$. Generalized inverse and some properties. Idempotent matrices and generalization of Graybill-Marsaglia results. Relationship between GI and idempotent matrices. MLE and combined estimates of β. Distributions. Goldberger-Theil-Kabe results. SCB and tests for $C\beta L$. Tests for second hypothesis. Largest root of a matrix.

Khatri, C.G. (1968): See Jaiswal, M.C. and Khatri, C.G., #502.

609. Khatri, C.G. and Pillai, K.C.S. (1968): On the noncentral distributions of two test criteria in multivariate analysis of variance. Ann. Math. Statist., 39, 215-226.

Evaluation of $\int|z|^{\nu}\alpha_p(z)c_\kappa(z)dz_1\ldots dz_{p-1}$ over the region $0 < z_1 < z_2 < \ldots < z_{p-1} < z_p = 1 - z_1 - \ldots - z_{p-1}$. Here z is diagonal $(z_1,\ldots,z_p)$. The density of $V^{(p)}$ in zonal polynomials. Moments in the linear case.

Tabulation of the coefficient $g^{\kappa}_{\alpha,\beta}$ and $b_{\kappa,\eta}$. Elementary symmetric functions of sub-matrices of L. Moments. Distribution of the largest root of L. Zonal polynomial representation. Tables.

610. Khatri, C.G. and Pillai, K.C.S. (1968): On the moments of elementary symmetric functions of the roots of two matrices and appoximations to a distribution. Ann. Math. Statist., 39, 1274-1281.

Elementary symmetric functions of a partitioned matrix in terms of the partitions. Application to e.s.f. of the roots of MV beta matrix. Moments. Approximations to the distribution of $U^{(p)}$ in the most general case. Patnaik's method and its generalizations. Comparison of the approximations.

611. Khatri, C.G. (1969): Noncentral distributions of the ith largest characteristic roots of three matrices concerning complex multivariate normal populations. Ann. Inst. Statist. Math., 21, 23-32.

Non-null distributions of the roots in MANOVA, covariance, and canonical correlation models. Distribution of the ith largest root by integration. Evaluation of the integral of the type $\int c_\kappa(W)\,|W|^{n-m}\exp(-tr\ W)\,|w_i^{m-j}|^2\cdot dw_1,\ldots,dw_m$ in the complex case. Bivariate covariance model in detail.

612. Khatri, C.G. (1969): A note on some results on a generalized inverse of a matrix. J. Indian Statist. Assoc., 7, 38-45.

Some new properties of generalized inverses. Applications to least squares estimation.

613. Khatri, C.G. and Pillai, K.C.S. (1969): Distributions of vectors corresponding to the largest roots of three matrices. Multivariate Anal.-II, [Proc. 2nd Internat. Symp., Krishnaiah, ed.], 219-240.

Some transformations and integrals involving zonal polynomials. CVCLR in two cases (i) one sample: (a) $\mu = 0$, (b) $\Sigma = 1$, (c) $\mu \neq 0$ but of rank 1. μ can be random. (ii) two sample case: (a) $\mu = 0$, (b) $\Sigma_1 = \Sigma_2$, (c) the rank μ is 1. MANOVA and canonical correlation. Testing hypothetical principal vectors of Σ_1 in the field of Σ_2, i.e., $\Sigma_1 b_i = \lambda_i \Sigma_2 b_i$, $i = 1,2,\ldots,r$; $\lambda_1,\ldots,\lambda_r$ are the roots of $\Sigma_1\Sigma_2^{-1}$. LR criterion. Decomposition of the hypothesis and the statistic. Distribution theory.

614. Khatri, C.G. (1970): Further contributions to some inequalities for normal distributions and their applications to simultaneous confidence bounds. Ann. Inst. Statist. Math., 22, 451-458.

Symmetric and separately symmetric convex regions. MVN distribution and inequalities over such regions. Confidence bounds on mean and one-sided confidence bound on variance.

615. Khatri, C.G. (1970): On the moments of traces of two matrices in three situations for complex multivariate populations. Sankhya, A, 32, 65-80.

Zonal polynomials and Laguerre polynomials of matrix argument (complex). Relationship with Hermite polynomials. Integrals involving such polynomials. Orthogonality. Integration involving Hermite polynomials and their orthogonality. Expected value of Laguerre polynomial with random argument. Moments of $tr(S^{-1}XL\bar{X}')$, $tr(S^{-1}X\bar{X}')$ *and* $tr\, XL\bar{X}'(S+XL\bar{X}')^{-1}$. *Tables of Laguerre polynomials.*

616. Khatri, C.G. (1970): A note on Mitra's paper "A density free approach to the matrix variate beta distribution". Sankhya, A, 32, 311-318.

Definition and study of MV beta distribution. Some transformations and their Jacobians. $E[C_K(\theta U)]$ *and related results.*

617. Khatri, C.G. (1971): On characterization of gamma and multivariate normal distributions by solving some functional equations in vector variables. J. Multivariate Anal., 1, 70-89.

Solution of some functional equations in vector variables which are an extension of Khatri-Rao results. Characterization of MVN using conditional expectation. Matrix multiplication extending Kronecker product and its properties.

618. Khatri, C.G. (1971): Series representations of distributions of quadratic forms in the normal vectors and generalized variance. J. Multivariate Anal., 1, 199-214.

Null and non-null distribution of (i) a quadratic form in normal vectors, $S = XAX'$ *and (ii)* $|S|$, *its generalized variance. Two types of series considered: (i) combined power series - Wishart type, (ii) Laguerre type. Zonal and Laguerre polynomials of matrix argument. Inequalities and bounds. Mellin transforms.*

619. Khatri, C.G. and Srivastava, M.S. (1971): On exact non-null distributions of likelihood ratio criteria for sphericity test and equality of two covariance matrices. Sankhya, A, 201-206.

Non-null distribution of LR criteria for testing (i) $\Sigma = \sigma^2 I$ *with* σ *unknown and (ii)* $\Sigma_1 = \Sigma_2$. *Laplace transforms. Results in terms of Meijer's G-function and H-function.*

620. Khatri, C.G. (1972): On the exact finite series distribution of the smallest or the largest root of matrices in three situations. J. Multivariate Anal., 2, 201-207.

Cumulative distribution function of the largest or the smallest characteristic root of matrices in three cases: (i) two-sample problem (ii) MANOVA (iii) canonical correlation problem. Distribution in terms of finite series. Generating functions. Laguerre polynomials of matrix argument.

621. Khatri, C.G. (1972): On the characterization of the distribution from the regression of sample covariance matrix on the sample mean vector. Sankhya, A, 34, 235-242.

Characterization based on the distribution of the sample covariance and its conditional expectation given the mean: $E(s_{ij}|\bar{x}) = a_{ij} + B_i\bar{x}_j + B_j\bar{x}_i + n\gamma_{ij}\bar{x}_j\bar{x}_k$. *Set of simultaneous partial differential equations. Multivariate Poisson, multinomial, negative multinomial, multivariate normal, multivariate gamma.*

Kiefer, J. (1963): See Giri, N., Kiefer, J. and Stein, C., #359.

Kiefer, J. (1964): See Giri, N. and Kiefer, J., #361.

Kiefer, J. (1964): See Giri, N. and Kiefer, J., #362.

622. Kiefer, J. and Schwartz, R. (1965): Admissible Bayes character of T^2-, R^2- , and other fully invariant tests for classical multivariate normal problems. Ann. Math. Statist., 36, 747-770. [Correction: 43 (1972), 1742].

Admissible Bayes procedures in multivariate normal testing problems. Consider: (i) MV general linear hypothesis, (ii) testing independence of sets of variates, (iii) tests concerning some component of mean vector, (iv) classification procedures, (v) Behrens-Fisher problem, (vi) equality of covariance matrices. Technique developed for obtaining Bayes procedures relative to a priori distributions of a certain type of problems where nuisance parameter means have been deleted.

623. King, B. (1967): Stepwise clustering procedures. J. Amer. Statist. Assoc., 62, 86-101.

Data analysis. Stepwise procedure using (i) maximization of pairwise correlation between the centroids of two groups, (ii) maximization of Wilks' criterion of independence between two groups. Applications.

624. Kingman, A. and Graybill, F.A. (1970): A nonlinear characterization of the normal distribution. Ann. Math. Statist., 41, 1889-1895.

Generalization of Skitovich's characterization of normality to bivariate and MV cases. Coefficients of the 'linear' functions are random variables.

625. Klatzky, S.R. and Hodge, R.W. (1971): A canonical correlation analysis of occupational mobility. J. Amer. Statist. Assoc., 66, 16-22.

Applications of canonical analysis. Assignment of weights to the various categories of occupational mobility tables. Use of canonical correlations between sets of dummy variables.

Klein, D.F. (1969): See Feldman, S., Klein, D.F. and Honigfeld, G., #297.

626. Klotz, J. and Putter, J. (1969): Maximum likelihood estimation of multivariate covariance components for the balanced one-way layout. Ann. Math. Statist., 40, 1100-1105.

Modification of the definition of MLE to include the case when matrices are not positive definite. Complete sufficient statistics. MLE of components. Computation. Bivariate case.

627. Koch, G.G. (1969): Some aspects of the statistical analysis of "split plot" experiments in completely randomized layouts. J. Amer. Statist. Assoc., 64, 485-505.

Parametric and nonparametric approaches to analysis of such experiments. Stress is on nonparametric aspects. Testing hypotheses of no treatment effect, no conditions effect, no interaction. Cole-Grizzle techniques. Compound symmetry of error distributions. Tests of the hypothesis concerning this assumption. Formulations of the models to fit the assumptions made.

628. Koch, G.G. and Lemeshow, S. (1972): An application of multivariate analysis to complex sample survey data. J. Amer. Statist. Assoc., 67, 780-782.

Comparison between the univariate and multivariate techniques bringing out the importance of the latter. No significant differences in mean heights (weights) found between six year old Negro and white children unless bivariate techniques used.

629. Korin, B.P. (1968): On the distribution of a statistic used for testing a covariance matrix. Biometrika, 55, 171-178.

LR criterion for testing H_0: $\Sigma = \Sigma_0$. Characteristic function and Box's expansion for the same. Distribution function in the null case. Approximations. Determination of significance points computed by various techniques. Tables of 5 and 1 percent points.

630. Korin, B.P. (1969): On testing the equality of k covariance matrices. Biometrika, 56, 216-218.

Comparison of significance points for M-tests of H_0: $\Sigma_1 = \ldots = \Sigma_k$. Series solution compared with χ^2 and F approximations.

631. Korin, B.P. (1971): Some comments on the homoscedasticity criterion M and the multivariate analysis of variance tests T^2, W and R. Biometrika, 58, 215-216.

Power of M and robustness of T^2, W and R criteria to heteroscedasticity. Tables of values of power and levels of significances: $k = 3, 6$.

632. Kotlarski, I. (1964): On bivariate random variables where the quotient of their coordinates follow some known distribution. Ann. Math. Statist., 35, 1673-1684.

Characterization of some bivariate distributions by the use of Mellin transforms.

633. Kotlarski, I. (1965): On the generalized Mellin transform of a complex random variable and its applications. Ann. Math. Statist., 36, 1459-1467.

Mellin transform of a complex random variable and inversion. Applications to complex normal distribution.

634. Kotz, S. and Adams, J.W. (1964): Distribution of sum of identically distributed exponentially correlated gamma variables. Ann. Math. Statist., 35, 277-283. [Correction: 35 (1964), 925].

Distribution of sum of correlated gamma variates when correlation is an exponential autocorrelation $\rho_{k,j} = \rho^{|k-j|}$, $k,j = 1,2,\ldots,n$. Approximation by a gamma distribution. Comparison with the exact distribution for certain values of ρ.

Kovats, M. (1966): See Fraser, A.R. and Kovats, M., #309.

635. Kowal, R.R. (1971): Disadvantages of the generalized variance as a measure of variability. Biometrics, 27, 213-216.

A criticism of the use of generalized variance as a measure of overall variability in MV populations. The correlation structure and its effects. Problem of estimation and estimated variance matrices. Precision.

636. Kowalski, C.J. and Tarter, M.E. (1969): Coordinate transformations to normality and the power of normal tests for independence. Biometrika, 56, 139-148.

Tests for independence after transformation to ensure marginal normality. Transformation of non-normal variates by normal distribution function. Fourier estimator and estimation of the transformation. Comparison of correlation coefficients obtained from the untransformed, transformed and by use of Fourier estimation procedures. Closeness of the probability densities to normal correlation density when $\rho = 0$. *Power of the tests of independence. More powerful when applied to transformed data. Several applications. Discussion.*

637. Kowalski, C.J. (1970): The performance of some rough tests for bivariate normality before and after coordinate trnasformations to normality. Technometrics, 12, 517-544.

Tests for bivariate normality: Quadrant test, Hald's line test, ring test, tests based on cumulants, Rosenblatt's transformation tests, MV Kolmogoroff-Smirnov and Cramer-von Mises tests. Transformations to marginal normality. Improvement to bivariate normality afforded by such transformations. Discussion and graphical display of results. Bibliographic.

638. Kraft, C.H., Olkin, I. and Van Eeden, C. (1972): Estimation and testing for differences in magnitude or displacement in the mean vectors of two multivariate normal populations. Ann. Math. Statist., 43, 455-467.

Tests of hypothesis on means of two MVN populations. H_1: $\mu = c\nu$ *against* H_2: $\mu = c\nu + d$. *MLE of* c, ν *and* Σ, *when* $d = 0$ *and of* c, ν, d *and* Σ *when* $d \neq 0$. *Here* c *and* d *are scalars. Unequal sample sizes. Asymptotic distributions of MLE. LR tests and asymptotic non-null distributions.*

639. Kramer, K.H. (1963): Tables for constructing confidence limits on the multiple correlation coefficient. J. Amer. Statist. Assoc., 58, 1082-1085.

Tables for values of sample multiple correlation coefficient when ρ (population coefficient) is .1(.1).9 and indices are 2a = 6 to 40, 2b = 10, 20, 30.

640. Krishnaiah, P.R. and Rao, M.M. (1961): Remarks on a multivariate gamma distribution. Amer. Math. Monthly, 68, 342-346.

Development of the MV gamma from the diagonal elements of a Wishart. Relationship between correlation structures. Infinitely divisible distributions. Characteristic function and its justification.

641. Krishnaiah, P.R., Hagis, P., Jr., and Steinberg, L. (1963): A note on the bivariate chi distribution. SIAM Rev., 5, 140-144.

Properties of the bivariate chi distribution. Moments and the conditional distribution. Distribution of sum of two correlated chi variates. Moments of the distribution of the ratio of two correlated chi variates. Applications.

642. Krishnaiah, P.R. (1964): Multiple comparison tests in multivariate case. ARL, 64-124, Wright Patterson AFB, Ohio.

Simultaneous procedures in MANOVA. Reviews several procedures: (i) MANOVA test based on largest root, (ii) T^2 (max) test, (iii) step-down procedure. Alternative procedure: finite intersection test. Distribution problems associated with finite intersection test. Simultaneous confidence intervals. Comparison of procedures. Results for both known and unknown covariance matrix.

643. Krishnaiah, P.R. (1965): On the simultaneous ANOVA and MANOVA tests. Ann. Inst. Statist. Math., 17, 35-53.

Simultaneous confidence intervals in MANOVA when hypotheses can be tested in a "quasi-independent" manner. Univariate case: confidence intervals when joint distribution sum of squares is multivariate chi-square. Comparison of intervals in terms of length. Procedures referred to as SANOVA and SMANOVA.

644. Krishnaiah, P.R. (1965): On a multivariate generalization of the simultaneous analysis of variance test. Ann. Inst. Statist. Math., 17, 167-173.

Multivariate generalization of SANOVA (see #643) test when several orthogonal hypotheses are to be tested simultaneously. Confidence intervals associated with procedures. Comparison with confidence intervals obtained from step-down procedures.

645. Krishnaiah, P.R. (1965): Simultaneous tests for the equality of variances against certain alternatives. Austral. J. Statist., 7, 105-109.

Problems concerned with rejection of H_0: $\sigma_1^2 = \sigma_2^2 = \ldots \sigma_p^2 = \sigma_0^2$. Procedures for comparing each variance simultaneously against next one. Union intersection principle. Simultaneous confidence intervals.

646. Krishnaiah, P.R. (1965): Multiple comparison tests in multi-response experiments. Sankhya, A, 27, 65-72.

Multiple comparisons of mean vectors against one-sided and two-sided alternatives. Σ unknown and known. Finite intersection tests: expressing total hypothesis as finite intersection of the elementary hypotheses concerning contrasts. Multivariate t, χ^2 and F distributions. Comparison with (i) MANOVA tests based on largest characteristic roots, (ii) test based on largest Hotelling T^2, (iii) step-down procedure. Simultaneous confidence bounds.

647. Krishnaiah, P.R. and Armitage, J.V. (1965): Tables for the distribution of the maximum of correlated chi-square variates with one degree of freedom. Trabajos Estadist., 16, 91-115.

Joint distribution of $(z_1, z_2, \ldots, z_p)$: p-variate (correlated) χ^2 with one degree of freedom. Distribution of maximum z. Application to a simultaneous test procedure, ranking and selection problems. Example of simultaneous testing procedures. Tables of upper 10, 5, 2.5 and 1 percentage points of the distribution of maximum z in the equicorrelated case for p = 1(1)10.

648. Krishnaiah, P.R., Hagis, P., Jr., and Steinberg, L. (1965): Tests for the equality of standard deviations in a bivariate normal population. Trabajos Estadist., 16, 3-15.

Testing equality of variances (standard deviations) against one-sided and two-sided alternatives. Samples from BVN with known correlation coefficient and unknown variances. Application to testing homogeneity of interactions in ANOVA. Tables.

649. Krishnaiah, P.R. and Armitage, J.V. (1966): Tables for multivariate t distribution. Sankhya, B, 28, 31-56.

Multivariate t distribution. Common unknown variance σ^2 *and known correlation matrix. Application to simultaneous testing procedures and ranking and selection problems. Tables of upper 1 and 5 percentage points in equicorrelated case with* $p = 1(1)10$.

650. Krishnaiah, P.R. and Rizvi, M.H. (1966): Some procedures for selection of multivariate normal populations better than a control. Multivariate Anal.-I, [Proc. 1st Internat. Symp., Krishnaiah, ed.], 477-490.

Selection procedures in MVN using (i) linear combinations of mean vectors: one-sided and two-sided cases, (ii) absolute values of linear compounds of the elements of the mean vectors, (iii) distance from the control, *(iv)* $\mu_0'\Sigma^{-1}\mu_0$ *and* $\mu_i'\Sigma^{-1}\mu_i$.

651. Krishnaiah, P.R. (1967): Selection procedures based on covariance matrices of multivariate normal populations. Trabajos Estadist., 18, 11-24.

Selection procedures based on covariance matrices using (i) diagonal elements, (ii) finite intersection of linear combinations of elements, (iii) all linear combinations of the elements, (iv) determinants, (v) largest or smallest roots. Bonferroni inequality and MV χ^2. *Probability of correct selection.*

652. Krishnaiah, P.R. and Pathak, P.K. (1967): Tests for the equality of covariance matrices under the intra-class correlation model. Ann. Math. Statist., 38, 1286-1288. [Correction: 39 (1968), 1358].

Simultaneous diagonalization of several covariance matrices, satisfying $\Sigma(\theta_1) = \Sigma(\theta_2)$ *if and only if* $\theta_1 = \theta_2$. *Covariance structures and models satisfying this characteristic. Test for the homogeneity of covariance matrices* Σ_i *where* $\Sigma_i = \sigma_i^2[(1-\rho_i)I+\rho_i J]$. *Si-*

multaneous diagonalization to diagonal $(\alpha_i, \beta_i, \beta_i, \ldots, \beta_i)$. *Test of* H_0: $\alpha_1 = \ldots = \alpha_k$, $\beta_1 = \ldots = \beta_k$ *by Roy's UI principal.*

653. Krishnaiah, P.R. (1968): Simultaneous tests for the equality of covariance matrices against certain alternatives. Ann. Math. Statist., 39, 1303-1309.

Stepwise procedure for testing the equality of several variance matrices. Split up of the hypothesis into subhypotheses. Confidence intervals. Union-intersection principle. Multivariate F distribution. Bonferroni's inequality. Comparison with Roy's conditional distribution method.

654. Krishnaiah, P.R. and Pathak, P.K. (1968): A note on confidence bounds for certain ratios of characteristic roots of covariance matrices. Austral. J. Statist., 10, 116-119.

Simultaneous CI on (i) $\lambda_{i+1,p}/\lambda_{i1}$ *and* $\lambda_{i+1,1}/\lambda_{ip}$ $(i = 1,2,\ldots,k-1)$, *(ii)* $\lambda_{i1}/\lambda_{jp}$ *and* $\lambda_{ip}/\lambda_{j1}$ $(i \neq j = 1,2,\ldots,k)$. *Here* λ_{ij} *is the jth largest root of* Σ_i $(j = 1,2,\ldots,k;\ i = 1,2,\ldots,k)$. *Khatri's method.*

655. Krishnaiah, P.V. (1969): Simultaneous test procedures under general MANOVA models. Multivariate Anal.-II, [Proc. 2nd Internat. Symp., Krishnaiah, ed.], 121-144.

Simultaneous test procedures under MANOVA, Potthoff-Roy growth curve model. Setting up T^2_{max}, *step-down and finite intersection confidence intervals. Comparison of the procedures in terms of confidence intervals and power. Behaviour as the sub-hypotheses increase. MANOVA model with autocorrelated errors and a general correlation pattern.*

656. Krishnaiah, P.R. and Waikar, V.G. (1969): Simultaneous tests for equality of latent roots against certain alternatives - II. ARL 69-0178, Wright Patterson AFB, Ohio.

Test of the hypothesis of equality of the latent roots of a matrix against the alternative that at least two are not equal. Random matrices considered: (i) Wishart matrix, (ii) $S_1S_2^{-1}$, *(iii) MANOVA matrix, (iv) canonical correlation matrix. Distribution of the ratio of largest root to the smallest root for each of the above matrices. Non-null case. Zonal polynomials and special functions of matrix argument.*

657. Krishnaiah, P.R., Armitage, J.V. and Breiter, M.C. (1969): Tables for the probability integrals of the bivariate t distribution. ARL 69-0060, Wright Patterson AFB, Ohio.

Tables of $(1-\alpha)$ for $n = 5(1)35$, $|\rho| = 0(0.1)0.9$, $a = 1.0(0.1)5.5$ where $(1-\alpha) = P\{t_1 \leq a,\ t_2 \leq a\}$. Five decimals.

658. Krishnaiah, P.R., Armitage, J.V. and Breiter, M.C. (1969): Tables for the bivariate $|t|$ distribution. ARL 69-0210, Wright Patterson AFB, Ohio.

Tables of $(1-\alpha)$ for $n = 5(1)35$, $|\rho| = 0(0.1)0.9$, $a = 1.0(0.1)5.5$, where $1-\alpha = P\{|t_1| \leq a,\ |t_2| \leq a\}$. Five decimals.

659. Krishnaiah, P.R. and Armitage, J.V. (1970): On a multivariate F distribution. Essays Prob. Statist., (Roy Volume, Bose et al, eds.), 439-468.

Bivariate F-distribution. Integral and approximations for $p > 2$. Distribution function of the maximum of multivariate t^2 variates. Tables of upper 5 and 1% points of t^2.

660. Krishnaiah, P.R. and Waikar, V.B. (1970): Exact distributions of the intermediate roots of a class of random matrices. ARL 70-0280, Wright Patterson AFB, Ohio.

Central distribution of rth root of a random matrix. Use of Laplace expansion of Vandermonde determinant. Recursive relation for the distribution function. Special cases of the matrices: Wishart, MANOVA, canonical correlation and one in nuclear physics.

661. Krishnaiah, P.R. and Chang, T.C. (1971): On the exact distribution of the smallest root of the Wishart matrix using zonal polynomials. Ann. Inst. Statist. Math., 23, 293-295.

An alternative expression for the density of the smallest root of the Wishart matrix. Zonal polynomials and hypergeometric function representation.

662. Krishnaiah, P.R. and Chang, T.C. (1971): On the exact distributions of the extreme roots of the Wishart and MANOVA matrices. J. Multivariate Anal., 1, 108-117.

Exact central distributions of the roots of the Wishart and MANOVA matrices. Laplace's expansion of Vandermonde's determinant. Joint distribution of the extreme roots. Extreme roots and their marginals.

663. Krishnaiah, P.R. and Waikar, V.B. (1971): Simultaneous tests for equality of latent roots against certain alternatives - I. Ann. Inst. Statist. Math., 23, 451-468.

Tests for equality of roots of certain random matrices. Subhypotheses and alternatives. Simultaneous testing procedures. Distributions of ratios of successive roots and of each root to the maximum root. Joint p.d.f. Application to Wishart, $S_1S_2^{-1}$, MANOVA, and canonical correlation matrices.

664. Krishnaiah, P.R. and Waikar, V.B. (1971): Exact distributions of any few ordered roots of a class of random matrices. J. Multivariate Anal., 1, 308-315.

Central distributions of a subset of ordered roots in MANOVA, Wishart and canonical correlation matrices. Pairs of ordered roots. Any intermediate root. Vandermonde determinant and its expansion.

665. Krishnaiah, P.R. and Chang, T.C. (1972): On the exact distributions of the traces of $S_1(S_1+S_2)^{-1}$ and $S_1S_2^{-1}$. Sankhya, A, 34, 153-160.

Central distribution of tr $S_1S_2^{-1}$ and tr $S_1(S_1+S_2)^{-1}$. Expansion of a pfaffian. Laplace transforms and their inversion. General case discussed for any p and m.

Krishnaiah, P.R. (1972): See Waikar, V.B., Chang, T.C. and Krishnaiah, P.R., #1151.

666. Krishnan, M. (1967): The noncentral bivariate chi distribution. SIAM Rev., 9, 708-714.

Noncentral χ in the bivariate case. Moments, conditional distribution and its moments. Sum and ratio of two noncentral χ variates. Applications.

667. Krishnan, M. (1972): Series representations of a bivariate singly noncentral t-distribution. J. Amer. Statist. Assoc., 67, 228-231.

Probability density and distribution function of the central and noncentral bivariate t. Series representations. Examples.

668. Kshirasagar, A.M. (1959): Bartlett decomposition and Wishart distribution. Ann. Math. Statist., 30, 239-241.

An alternative demonstration of Bartlett decomposition of the Wishart distribution using random orthogonal transformations. Explicit expressions for the χ^2 and the normal variables.

669. Kshirasagar, A.M. (1960): A note on the derivation of some exact multivariate tests. Biometrika, 47, 480-482.

Derivation of results stated in Williams' paper on discriminant and canonical analyses. Direct approach. Joint distribution of W and D when KAK' = WDW' and K(A+B)K' = WW'. Central and noncentral cases when one canonical correlation is nonzero (linear case). True discriminant and the residual after the elimination of the linear relation between the true discriminant and the y vector. Linkage factor.

670. Kshirasagar, A.M. (1961): The noncentral multivariate beta distribution. Ann. Math. Statist., 32, 104-111.

The noncentral MV beta is defined for the linear and planar cases. Application to the decomposition of Wilks' Λ criterion in the canonical correlation case. Representation of t_{ii}^2 as a function of ratio of multiple correlations. Simplification in the linear case. Conditional distribution of the two canonical correlation coefficients. Discriminant analysis.

671. Kshirasagar, A.M. (1961): The goodness-of-fit of a single (non-isotropic) hypothetical principal component. Biometrika, 48, 397-407.

Overall χ^2 test of goodness-of-fit of an hypothetical non-isotropic principal component. Split up of the degrees of freedom into components due to (i) there being more than one true non-isotropic principal component, (ii) the hypothetical component not being the same as the true (directional difference). Expression for the directional component of the χ^2 in (i) rectangular coordinates, (ii) terms of sample latent roots and 'residual' roots. Relationship with Williams' discriminant analysis. Conditional distributions. Example.

672. Kshirasagar, A.M. (1961): Some extensions of the multivariate t-distribution and the multivariate generalization of the distribution of the regression coefficient. Proc. Cambridge Philos. Soc., 57, 80-85.

Noncentral generalization of the MV t. Selection of the population with largest mean. Distribution of $\{t_{ii}\}$ in noncentral (linear) case when $\Sigma = \sigma^2 R$, R known and σ^2 unknown. Regression coefficient and its MV analogue in the case of $\mu = 0$. Extension of Cornish's results. Derivation of the distribution of the regression matrix. MV F distribution.

673. Kshirasagar, A.M. (1962): Effect of non-centrality on the Bartlett decomposition of a Wishart matrix. Ann. Inst. Statist. Math., 14, 217-228.

The effect of noncentrality on the elements of the component matrix T in W = TT'. Random orthogonal transformation. Consideration of (i) linear, (ii) planar, (iii) three nonzero roots cases. The differences between the various cases. Complication of dependencies in the planar case.

674. Kshirasagar, A.M. (1962): A note on direction and collinearity factors in canonical analysis. Biometrika, 49, 255-259.

Anomalies between Bartlett and William discriminant analyses of Barnard's anthropometric data. Discriminant analysis of (q+1) groups with k variables. ANOVA of the regression. Wilks' Λ criterion and its representation as $(1-\phi)\Lambda^1$, where Λ^1 is the residual criterion after elimination of the dependence of X on the hypothetical function. Further factorizations and their interpretations as the collinearity and directional factors. Example.

675. Kshirasagar, A.M. (1963): Confidence intervals for discriminant function coefficients. J. Indian Statist. Assoc., 1, 1-7.

Discriminant analysis in multiple groups viewed as regression analysis. Relation between canonical analysis (linear case) and discriminant analysis. Tests of significance and confidence intervals in the case of collinearity (single discriminant function). Tests for $\alpha_1/\alpha_p = k$, a specified quantity; here $\alpha_1, \alpha_2, \ldots, \alpha_p$ are the coefficients of the single discriminant function. Confidence intervals on the ratios of the coefficients using the direction factor. Approximate methods for the non-collinear case. Example.

676. Kshirasagar, A.M. (1964): Wilks' Λ criterion. J. Indian Statist. Assoc., 2, 1-20.

Expository article that lists all the properties of Wilks' Λ. Its decomposition and applications are discussed. Bibliographic.

677. Kshirasagar, A.M. (1964): Distributions of the direction and collinearity factors in discriminant analysis. Proc. Cambridge Philos. Soc., 60, 217-225.

Wilks' Λ criterion and its null distribution. Collinearity, tests of equality of means, adequacy of a single linear discriminant function for q groups. Factorization of Λ into factors Λ_1, Λ_2 (direction), Λ_3 (partial collinearity). Distributional properties of the factors by associating them with the elements of T where $\Lambda = TT'$. Alternative factorizations of Λ into factors Λ_1, Λ_4 (collinearity), Λ_5 (partial direction). Distributions of the factors. All distribution theory is in the central case.

678. Kshirasagar, A.M. and Gupta, R.P. (1965): The goodness-of-fit of two (or more) hypothetical principal components. Ann. Inst. Statist. Math., 17, 347-356.

Extension of Kshirasagar's results to two or more hypothetical principal components. Decomposition of the overall χ^2 into components due to each of the hypothesized principal components. Directional and angular components. Generalizations and example.

679. Kshirasagar, A.M. and Gupta, R.P. (1965): A note on the use of canonical analysis in factorial experiments. J. Indian Statist. Assoc., 3, 165-169.

Analysis of the factorial (proportionality) model $x_{it} = f_i g_t + e_{it}$ as a canonical analysis. Single non-isotropic principal component.. Distribution of test statistics. Relationship with Kshirasagar's results.

680. Kshirasagar, A.M. (1966): The non-null distribution of a statistic in principal component analysis. Biometrika, 53, 590-594.

The statistic χ_d^2 used for testing H_0: $\gamma = \gamma_1$ in principal component analysis, its representation and distributions. Conditional on the hypothetical vector and unconditional distributions.

681. Kshirasagar, A.M. (1969): Distributions associated with the factors of Wilks' Λ in discriminant analysis. J. Austral. Math. Soc., 10, 269-277.

Relationship between discriminant analysis and Wilks' Λ. Goodness-of-fit of s hypothetical discriminants. Canonical correlations. Factorization of Λ into various factors. Representation in terms of elements of the T matrix (lower triangular). Distributions of the factors.

682. Kshirasagar, A.M. (1969): Correlation between two vector variables. J. Roy. Statist. Soc., B, 31, 477-485.

Generalization of the correlation coefficient r to matrix form R. Ruben's approximation to $r/(1-r^2)^{\frac{1}{2}}$ and its matrix analogues. Non-null distributions. Direction and collinearity factors of the Hotelling-Lawley and Pillai criteria. Large sample tests corresponding to the partitions.

683. Kshirasagar, A.M. (1970): Alternative derivation of the distribution of the direction and collinearity statistics in discriminant analysis. Calcutta Statist. Assoc. Bull., 18, 123-134.

Representations for the direction and collinearity factors in terms of Wishart matrices. Distributions. Goodness-of-fit of a single and of s hypothetical discriminant functions.

684. Kshirasagar, A.M. (1971): Goodness-of-fit of a discriminant function from the vector space of dummy variables. J. Roy. Statist. Soc., B, 33, 111-116.

In the discriminant analysis using dummy variables, the representation of Wilks' Λ alternatively in terms of the dummy variables. Duality in the null case. Distributions of the factors of the Λ criterion under the null hypothesis that one discriminant function in the space of the dummy variables is adequate. Collinearity and directional aspects. Wijsman's random transformations technique. Generalizations to more than one hypothetical function.

685. Kshirasagar, A.M. (1971): Review of the Williams-Bartlett-Kshirasagar work on direction and collinearity factors in discriminant analysis. Tech. Rep. #107, Dept. Statist., Southern Methodist U.

Bibliographic article on discriminant analysis involving tests on direction and collinearity factors. Tests on several discriminant functions.

686. Kudo, A. (1957): The extreme value in a multivariate normal sample. Mem. Fac. Sci. Kyushu Univ., Ser. A, 11, 143-156.

Tests of an outlier in terms of the mean vector. $X^{(\nu)}$ is the observation vector with known variance matrix Σ. H_0: $m^{(\nu)} = m$ is the null, with H_ν specifying that $X^{(\nu)}$ only has a mean vector differing from the others. D_ν $(\nu = 0,1,\ldots,n)$, the decision to accept H_ν subject to certain criteria. Accept H_k if the value of k is such that T_k^2 is the maximum of $T_1^2, T_2^2,\ldots,T_n^2$ and $T_k^2 \geq c_\alpha$. Here T_ν^2 is the quadratic form $\Sigma\Sigma(x_i^{(\nu)}-\bar{x}_i)(x_j^{(\nu)}-\bar{x}_j)\sigma^{ij}$ and c_α is chosen so that the probability of a correct decision when H_0 is true is $(1-\alpha)$.

687. Kudo, A. (1958): On the distribution of the maximum value of an equally correlated sample from a normal population. Sankhya, 20, 309-316.

Distribution function of a MVN when $\mu = 0$ *and* $\Sigma = \sigma^2[I(1-\rho) + \rho J]$. *Distribution function of the* $\max\{x_1, x_2, \ldots, x_p\}$, *i.e., pth order statistic for p correlated normal variables. Tables.*

688. Kudo, A. (1959): The classificatory problem viewed as a two decision problem. Mem. Fac. Sci. Kyushu Univ., Ser. A, 13, 96-125.

Classification of an observation into one of k populations. Vector-valued classification function. Power and generalized power. Invariance and sufficient statistics. MVN with known means and known (unknown) covariances. Unequal sample sizes. Optimal within the class of location (location and scale) invariant procedures with power functions satisfying certain symmetry conditions.

689. Kudo, A. (1960): The classificatory problem viewed as a two decision problem, II. Mem. Fac. Sci. Kyushu Univ., Ser. A, 14, 63-83.

Stringency and most stringent classification rules. Symmetry relative to distance function on the parameter space. Proof that the standard classification procedure with μ *unknown,* Σ *known or unknown is most stringent. Modifications of the theorems of Riesz and of Hunt and Stein.*

690. Kudo, A. (1961): Some problems of symmetric multiple decisions in multivariate analysis. Bull. Inst. Internat. Statist., 38, 165-171.

Application to (i) determination of extreme values in MVN, (ii) classification using linear discriminant function. Symmetric multiple decision problems and most stringent functions.

Kudo, A. (1962): See Schull, W.J. and Kudo, A., #1006.

691. Kudo, A. (1963): A multivariate analogue of the one-sided test. Biometrika, 50, 403-418.

X is MVN with $E(X) = \theta$, $Var(X) = \Sigma$ *(known). Test of the hypothesis* H_0: $\theta = 0$ *against the restricted alternative* H_A: $\theta > 0$. *LR test and the distribution of the criterion.*

692. Kudo, A. and Fujisawa, H. (1966): Some multivariate tests with restricted alternative hypotheses. Multivariate Anal.-I, [Proc. 1st Internat. Symp., Krishnaiah, ed.], 73-85.

One-sided test of the mean in BVN: H_0: $\theta_1 = 0$, $\theta_2 = 0$ *against* H_A: $\theta_1 \leq 0$, $\theta_2 \leq 0$ *or* $\theta_1 \geq 0$, $\theta_2 \geq 0$, *when at least one of the inequalities (in* H_A*) is strict in both cases. Power of the LR test. Tables of the power. MV generalizations when* Σ *is known.*

693. Kullback, S. and Rosenblatt, H.M. (1957): On the analysis of multiple regression in k categories. Biometrika, 44, 67-83.

MV multiple linear hypothesis. Tests of hypotheses and sub-hypotheses. Information theoretic approach.

694. Kullback, S. (1967): On testing correlation matrices. Appl. Statist., 16, 80-85.

Numerical examples of the various test procedures for (i) H_0: $R = P$ *(prescribed), (ii)* H_0: $R_1 = R_2 = \ldots = R_k$, *equality of several correlation matrices. Sub-hypotheses concerning the independence of one variate with the others. Split up of degrees of freedom.*

695. Kurczynski, T.W. (1970): Generalized distance and discrete variables. Biometrics, 26, 525-534.

The measures of distances between two or more populations when measurements are frequencies or proportions. Numerical evaluations of three such measures. Discussion.

696. Kurtz, T.E., Link, R.F., Tukey, J.W. and Wallace, D.L. (1966): Correlation of ranges of correlated deviates. Biometrika, 53, 191-197.

(X, Y) is BVN with zero means and standardized variance $U = $ *range* $\{X_1, \ldots, X_n\}$, $V = $ *range*$\{Y_1, \ldots, Y_n\}$. *Exact correlation between U and V for* $n = 2, 3$. *Asymptotic results for large n. Behaviour for* $\rho \to 0$.

Kuzmack, A. (1967): See Freeman, H., Kuzmack, A. and Maurice, R., #312.

Kuzmack, A.M. (1972): See Freeman, H. and Kuzmack, A.M., #313.

697. Kymn, K.O. (1968): The distribution of the sample correlation coefficient under the null hypothesis. Econometrica, 36, 187-189.

Distribution of $s = (1+r)/(1-r)$, *r is sample correlation, when* $\rho = 0$. *s is* $F(n-2, n-2)$.

698. Lachenbruch, P.A. (1966): Discriminant analysis when the initial samples are misclassified. Technometrics, 8, 657-662.

Fisher's linear discriminant function. Effects of estimating μ_1, μ_2, Σ with misclassified initial samples. Small sample results. Monte Carlo techniques. Large sample results.

699. Lachenbruch, P.A. (1967): An almost unbiased method of obtaining confidence intervals for the probability of misclassification in discriminant analysis. Biometrics, 23, 639-645.

Unbiased estimation of probabilities of misclassification using all of the data. Fisher's linear discriminant function. Confidence intervals. Sampling experiments and discussion.

700. Lachenbruch, P.A. (1968): On expected probabilities of misclassification in discriminant analysis, necessary sample size, and a relation with the multiple correlation coefficient. Biometrics, 24, 823-834.

Error rates and their expected values. Sample size required to obtain error rates within given tolerance limits: (i) large tolerances, (ii) widely separated groups, (iii) large number of parameters. Multiple correlation and effects of its 'shrinkage' on the probabilities of misclassification.

701. Lachenbruch, P.A. and Mickey, M.R. (1968): Estimation of error rates in discriminant analysis. Technometrics, 10, 1-11.

Estimation of probability of misclassification. Monte Carlo studies. Comparison of techniques: (i) Mahalanobis' D^2 method, (ii) D^2 corrected for bias, (iii) estimation based on expansion due to Okamoto, (iv) empirical procedures based on dividing the sample.

702. Ladd, G.W. (1966): Linear probability functions and discriminant functions. Econometrica, 34, 873-885.

Linear probability function: regression function in which the dependent variable has the value of zero or one. Relationships between linear probability functions and discriminant functions. Examples of discriminant analysis in economics.

703. Lancaster, H.O. (1957): Some properties of the bivariate normal distribution considered in the form of a contingency table. Biometrika, 44, 289-292.

Relationship between discriminant analysis and analysis of contingency tables. Maximal property of BVN: ρ, the correlation coefficient, is the maximum canonical correlation. Hermite-Tchebycheff's polynomials. Joint density expressed as Mehler series. Relationship between χ^2 and correlation.

704. Lancaster, H.O. (1958): The structure of bivariate distributions. Ann. Math. Statist., 29, 719-736. [Correction: 35 (1964), 1388].

Canonical correlation theory in terms of eigenfunction theory. Pearson's mean square contingency ϕ^2. ϕ^2 as sum of squares of the correlation coefficients. Properties of canonical variables. Regression in the bivariate distribution. Canonical partition of χ^2. Test for bivariate normality.

705. Lancaster, H.O. (1959): Zero correlations and independence. Austral. J. Statist., 1, 53-56.

Relationship between zero correlation and independence. Independence of two linear forms of MVN variables if and only if linear forms are uncorrelated. Necessary and sufficient condition for the mutual independence of x and y: every function of x with finite variance is uncorrelated with every function of y with finite variance. Independence in terms of canonical correlation.

706. Lancaster, H.O. (1960): The characterization of the normal distribution. J. Austral. Math. Soc., 1, 368-383.

Existence of moments of all orders in N(0, 1). Proof of well-known characterizations of the normal distribution directly, without use of cumulants.

707. Lancaster, H.O. (1960): On statistical independence and zero correlation in several dimensions. J. Austral. Math. Soc., 1, 492-496.

Generalizations of #704 and #705 to multivariate situation. Independence in terms of generalized coe-coefficients of correlation.

708. Lancaster, H.O. (1963): Correlations and canonical forms of bivariate distributions. Ann. Math. Statist., 34, 532-536.

Expansion for bivariate distributions in terms of marginal distributions and the canonical correlations. Characterization of bivariate distributions by correlation. Canonical correlations in ϕ^2-unbounded distributions.

709. Lancaster, H.O. (1963): Correlation and complete dependence of random variables. Ann. Math. Statist., 34, 1315-1321.

Concepts of complete independence and complete dependence. Complete dependence of k random variables. Examples of general situations in which measures of dependence used for normal distribution are of little interest.

710. Lancaster, H.O. (1965): The Helmert matrices. Amer. Math. Monthly, 72, 4-12.

Generalized Helmert matrix. Properties and factorization of Helmert matrices. Application in the evaluation of the Jacobian of the transformation from rectangular to polor coordinates in n-dimensions.

711. Lancaster, H.O. (1965): Symmetry in multivariate distributions. Austral. J. Statist., 7, 115-126.

Definition and test for symmetry in the bivariate case. Definition of symmetry in multivariate distributions in terms of generalized correlation coefficients. Testing for symmetry by partitioning χ^2.

712. Lancaster, H.O. (1966): Kolmogorov's remark on the Hotelling canonical correlations. Biometrika, 53, 585-588.

Alternative development of Hotelling's canonical correlations in the MVN. Canonical variables in a generalized sense. Maximum correlation. Proof of Kolomogorov's remark as a generalization of Lancaster (#703).

713. Laubscher, N.F. (1959): Note on Fisher's transformation of the correlation coefficient. J. Roy. Statist. Soc., B, 21, 409-419.

Class of transformations $Z = \frac{1}{2}\ln\{(c_1+r)/(c_2-r)\}$. Determine c_1 and c_2 such that Var Z is independent of ρ and Z is approximately normal. Optimality of Fisher's transformation ($c_1 = c_2 = 1$).

714. Lawley, D.N. (1959): Tests of significance in canonical analysis. Biometrika, 46, 59-66.

Moments of the canonical correlation coefficients. Approximate distribution of first k roots. Approximate test of significance for the residual canonical roots when the effect of the first k roots has been removed. Canonical analysis when one vector represents a set of fixed dependent variates: regression analysis.

715. Lawley, D.N. (1963): On testing a set of correlation coefficients for equality. Ann. Math. Statist., 34, 149-151.

Test of hypothesis: $\rho_{ij} = \rho$. *Generalization of Anderson's result for* $p = 3$. *Latent roots of correlation matrix. Asymptotic test criterion based on* χ^2.

716. Layard, M.W.J. (1972): Large sample test for the equality of two covariance matrices. Ann. Math. Statist., 43, 123-141.

Tests of $\Sigma_1 = \Sigma_2$. *Non-robust nature of usual tests: (i) LR criterion, (ii) Roy and Gnanadesikan test based on largest and smallest root of* $S_1S_2^{-1}$. *Asymptotically robust procedures: (i) test based on elementary symmetric functions of the roots of* $S_1S_2^{-1}$, *(ii) a "standard error test": vector of differences of transformed second-order sample moments is standardized by means of an estimate of asymptotic covariance matrix, (iii) test based on Box's idea of dividing data into groups and examining functions of second-order moments of each group, (iv) jack-knife procedure. Comparison of procedures using Pitman efficiency and approximate Bahadur efficiency.*

Lee, E.T. (1970): See Gnanadesikan, R. and Lee, E.T., #382.

717. Lee, J.C. and Geisser, S. (1972): Growth curve prediction. Tech. Rep. #167, School Statist., U. Minnesota.

Partial or conditional prediction in growth curve analysis when Σ *is assumed to have structure* $\Sigma = X\Gamma X' + Z\theta Z'$. *Predictive distributions. Prediction of a vector of future observations after having observed a portion of it. Application to estimation of "missing values". Estimation of* Σ. *LR test for* Σ *structured vs.* Σ *arbitrary.*

718. Lee, J.C. and Geisser, S. (1972): Applications of growth curve prediction. Tech. Rep. #180, School Statist., U. Minnesota.

Continuation of #717. Empirical study of various predictors for a variety of covariance models. Illustration of techniques using growth curve data.

719. Lee, Y.S. (1971): Distribution of the canonical correlations and asymptotic expansions for distributions of certain independence test statistics. Ann. Math. Statist., 42, 526-537.

Representation of sample canonical correlations as the roots of a determinantal equation involving independent matrix variates. Analogy with Roy's equation in the linear hypothesis case. Asymptotic expansion of an integral involving a characteristic function. Asymptotic expansions for the criteria: (i) Wilks' LR, (ii) Hotelling's T_0^2, (iii) Pillai's V using the expansion in the linear hypothesis case. Extension of Ito's expansion for T_0^2, in the linear hypothesis case, to terms of $O(n^{-2})$.

720. Lee, Y.S. (1971): Asymptotic formulae for the distribution of a multivariate test statistic: power comparisons of certain multivariate tests. Biometrika, 58, 647-651.

General linear hypothesis and test criteria: Wilks' LR, Hotelling's T_0^2, Pillai's V. Asymptotic expansion to $O(n^{-3})$ of Pillai's V. Upper percentage point for V by Hill-Davis method. Power comparisons. Tables of approximate powers.

721. Lee, Y.S. (1971): Some results on the sampling distribution of the multiple correlation coefficient. J. Roy. Statist. Soc., B, 33, 117-130.

Representation of $R^2/(1-R^2)$ in Ruben (simple correlation) form. Distribution function of R^2 and difference equations for the integral in case of odd p. Asymptotic expansions: (i) series of noncentral beta, (ii) noncentral χ^2, (iii) central and noncentral F using three moments. Normalizing by power transformations: choice of h. Accuracy of the approximations. Tables. Fisher's tan h^{-1} transformation and its invalidity. Editorial note concerning Fisher's transformations.

722. Lee, Y.S. (1972): Tables of upper percentage points of the multiple correlation coefficient. Biometrika, 59, 175-189.

Unconditional distribution of R^2. Computation of the upper percentage points using Newton-Raphson method. Interpolation methods. Approximations. Tables.

723. Lee, Y.S. (1972): Some results on the distribution of Wilks's likelihood-ratio criterion. Biometrika, 59, 649-664.

Null distribution of Wilks' Λ *when (i) p or q is even (ii) both are odd. Asymptotic expansions. Tables for* $p \leq q \leq 20$, $pq < 144$ *except when p or q is odd. Tables of the correction for* χ^2.

724. Lees, R.W. and Lord, F.M. (1961): A nomograph for computing partial correlation coefficients. J. Amer. Statist. Assoc., 56, 995-997. [Correction: 57 (1962), 917-918].

A nomograph for computing $r_{12.3}$. *Directions for the use of the tables. Example.*

Lemeshow, S. (1972): See Koch, G.G. and Lemeshow, S., #628.

Lewis, T. (1963): See Barnett, V.D. and Lewis, T., #76.

Li, C.C. (1966): See Degroot, M.H. and Li, C.C., #239.

Li, C.C. (1968): See Rao, B.R., Garg, M.L. and Li, C.C., #925.

Li, H.C. (1970): See Pillai, K.C.S. and Li, H.C., #899.

725. Li, H.C., Pillai, K.C.S. and Chang, T.C. (1970): Asymptotic expansions for distributions of the roots of two matrices from classical and complex Gaussian populations. Ann. Math. Statist., 41, 1541-1556.

Asymptotic expansion of an integral occurring in the joint distribution of the roots of S (one-sample case) and $S_1S_2^{-1}$ *(two-sample case). Expansion based on a representation of the determinant* $|I+AQBQ'|$. *Evaluation of its maximum with respect to Q. Extension of Chang's result in two-sample case. Complex analogues.*

Liao, S.H. (1970): See Patil, S.A. and Liao, S.H., #866.

726. Lin, P.E. (1971): Estimation procedures for difference of means with missing data. J. Amer. Statist. Assoc., 66, 634-636.

ML estimation of $\mu_1 - \mu_2$ *when sampling from a bivariate normal distribution with unknown* Σ. *Some observations missing in one variable. Comparison of estimators using MSE: Mehta-Gurland estimator and regression-type estimator.*

727. Lin, P.E. (1972): Some characterizations of the multivariate t distribution. J. Multivariate Anal., 2, 339-344.

Representations of the multivariate t distributions: (i) a normal vector and an independent χ^2 variable, (ii) a normal vector and an independent Wishart matrix. Characterization based on (i) linear functions of the vector of t-variables, (ii) relation to F-distribution, (iii) a property of the p.d.f. using a function of $x_1^2 + x_2^2 + \dots + x_p^2$.

728. Linhart, H. (1959): Techniques for discriminant analysis with discrete variables. Metrika, 2, 138-149.

Classification of an individual into one of two groups based on measurements made on discrete (or discrete and continuous) variables. Assumes joint distribution known. Develops classification rule which minimizes the expected cost of misclassification. Estimation of joint distribution when unknown. Comparison with traditional discriminant analysis for continuous variables.

Link, R.F. (1966): See Kurtz, T.E., Link, R.F., Tukey, J.W. and Wallace, D.L., #696.

Lipton, S. (1957): See Ashton, E.M., Healy, M.J.R. and Lipton, S., #51.

Livingstone, D. (1965): See Rayner, A.A. and Livingstone, D., #951.

729. Lockhart, R.S. (1967): The assumption of multivariate normality. British J. Math. Statist. Psychol., 20, 63-69.

Tests for multivariate normality based on marginal normality and transformation for independence. Ad hoc procedure. Example for BVN case.

730. Lord, F.M. (1955): Estimation of parameters from incomplete data. J. Amer. Statist. Assoc., 50, 870-876.

MLE in TVN distribution when (X, Z) and (Y, Z) are observed and no observations on (X, Y) are available. Efficiency of the estimators. Numerical example.

Lord, F.M. (1961): See Lees, R.W. and Lord, F.M., #724.

731. Lowe, J.R. (1960): A table of the integral of the bivariate normal distribution over an offset circle. J. Roy. Statist. Soc., B, 22, 177-187.

Probability content in a circle with arbitrary center and radius when (X, Y) are independent normals with zero means and variances σ_1^2, σ_2^2. *Tables of the areas.*

732. Lubischew, A.A. (1962): On the use of discriminant functions in taxonomy. Biometrics, 18, 455-477.

Consideration of scatter and correlation ellipses. Rank of discrimination. Effect of increase of number of variables on the rank. Use for more than two groups by pairwise determinations of the rank. Compound characters. Selection of single characters. Taxonomic implications.

733. Lyttkens, E. (1972): Regression aspects of cannoical correlation. J. Multivariate Anal., 2, 418-439.

Determination of canonical variables by regression. Iterative procedures.

734. MacNaughton-Smith, P. (1963): The classification of individuals by the possession of attributes associated with a criterion. Biometrics, 19, 364-366.

Classification with dichotomous observations. Criterion based on χ^2 *statistic for 2x2 table. Estimation of probability of success.*

735. MacQueen, J. (1967): Some methods for classification and analysis of multivariate observations. Proc. Fifth Berkeley Symp. Math. Statist. Prob., 1, 281-297.

Partitioning an n-dimensional population into k sets on the basis of a sample. Procedure called "k-means". Asymptotic behaviour of procedure. Discussion of situations where method is applicable.

736. Madansky, A. (1959): Bounds on the expectation of a convex function of a multivariate random variable. Ann. Math. Statist., 30, 743-746.

Upper and lower bounds on the expectation of a convex function of a vector valued random variable using the boundary of an appropriate moment space. Multidimensional Jensen's inequalities. Moment spaces. Moment inequalities.

737. Madansky, A. (1964): Spurious correlation due to deflating variables. Econometrica, 32, 652-655.

Multiple linear regression with random independent variables. Deflation. Approximation for $Cov(Y/Z, X/Z)$ when $Cov(X,Y) = 0$ and X, Y, Z are random variables. Discussion of spurious correlation. Fieller's theorem.

738. Madansky, A. and Olkin, I, (1969):: Approximate confidence regions for constraint parameters. Multivariate Anal.-II, [Proc. 2nd Internat. Symp., Krishnaiah, ed.], 261-288.

Confidence regions for covariance matrix and its functions in one and two sample case. Methods used: (i) direct linearization and its asymptotic distribution, (ii) LR statistic method using Lagrangian multipliers. Functions of Σ in one sample case: $tr\Sigma^{-1}A$, $tr\Sigma A$, $|\Sigma|$, characteristic roots of Σ. Functions of Σ_1, Σ_2 in two sample case: $tr(A_1\Sigma_1^{-1}+A_2\Sigma_2^{-1})$, $tr(A_1\Sigma_1+A_2\Sigma_2)$, $|\Sigma_1|^{a_1}|\Sigma_2|^{a_2}$, $a_1|\Sigma_1^{-1}|+a_2|\Sigma_2^{-1}|$, $tr\Sigma_1^{-1}\Sigma_2$. Behrens-Fisher problem.

Mahmoud, M.W. (1964): See Mostafa, M.D. and Mahmoud, M.W., #815.

739. Majumdar, D.N. and Rao, C.R. (1958): Bengal anthropometric survey, 1945: A statistical study. Sankhya, 19, 201-408.

Application of Mahalanobis' generalized distance to a study of anthropometric survey. Detailed data are given for 3240 individuals belonging to 41 social groups. Fourteen different characteristics are studied.

740. Mallows, C.L. (1961): Latent vectors of random symmetric matrices. Biometrika, 48, 133-149.

Latent vectors of $X = \Sigma+Z$, where Σ is fixed and Z is a random symmetric matrix with zero means. Covariance structures of Z: (i) invariant covariance, (ii) Wishart, (iii) rounding-off. LR criteria developed for testing the hypothesis that a specified set of p orthogonal unit vectors are the latent vectors of Σ. Moments of statistics: non-normal version of invariant structure and for rounding-off structure. Confidence regions and approximations.

741. Mandel, J. (1970): The distribution of eigenvalues of covariance matrices of residuals in analysis of variance. J. Res. Nat. Bur. Standards Sect. B, 74, 149-154.

Definition of "interaction matrix" of r degrees of freedom by s degrees of freedom. Partitioning "interaction matrix" in a two-way ANOVA problem using principal component analysis. Distribution of eigenvalues of "interaction matrix". Vacuum cleaner.

742. Mandel, J. (1971): A new analysis of variance model for non-additive data. Technometrics, 13, 1-18.

Use of principal component analysis to partition interaction sum of squares in two-way ANOVA. Several examples of technique. Calculation of expectation and variance for partitions using Monte Carlo techniques. Use in data analysis.

743. Mandel, J. (1972): Principal components, analysis of variance and data structure. Statistica Neerlandica, 26, 119-129.

Principal component analysis viewed as an analysis of variance. Interpretation of principal components. Data structure. Examples.

744. Maniya, G.M. (1968): Quadratic error in estimating multi-dimensional normal distribution densities from sample data. Theory Prob. Appl., 13, 341-343.

Estimation of a MVN density function with parameters $(0, \Sigma)$. Quadratic error $\int\{f(x)-f^*(x)\}^2 dx$, where $f(x)$ is the p.d.f. of X and $f^*(x)$ the value of $f(x)$ with Σ replaced by sample variance matrix S.

745. Maniya, G.M. (1969): The square error of the density estimate of a multi-dimensional normal distribution for a given sample. Theory Prob. Appl., 14, 149-153.

Estimation of the MVN density in the general (μ, Σ) case. Proof that in the limiting case the quadratic error, $\phi_n = \int[n(\mu, \Sigma)-n(\bar{x}, S)]^2 dx$, is reduced to considering the limiting form of $\psi_n = \int[n(0, I)-n(\bar{y}, T)]^2 dy$. Thus $\lim P\{c\phi_n < x\} = F(x)$, a linear function of chi-square distribution functions for a suitable choice of the constant c.

Mann, D.W. (1967): See Beale, E.M.L., Kendall, M.G. and Mann, D.W., #82.

Mantel, N. (1962): See Geisser, S. and Mantel, N., #335.

Mantel, N. (1963): See Halperin, M. and Mantel, N., #433.

746. Mantel, N. (1966): Corrected correlation coefficients when observation on one variable is restricted. Biometrics, 22, 182-187.

Correlation between X and Y when X is restricted (in observation) to lie in some interval. Relationship between the restricted and unrestricted moments. Expression for ρ in terms of ρ' (the restricted correlation coefficient).

Marcus, L.F. (1965): See Banerjee, K.S. and Marcus, L.F., #70.

Marcus, L.F. (1968): See Chaddha, R.L. and Marcus, L.F., #141.

747. Mardia, K.V. (1962): Multivariate Pareto distributions. Ann. Math. Statist., 33, 1008-1015. [Correction: 34 (1963), 1603].

Bivariate Pareto distributions of types I and II. Marginal, conditional distributions. Regression. Estimation of the parameters: ratio and regression methods. MV generalizations.

748. Mardia, K.V. (1964): Some results on the order statistics of the multivariate normal and Pareto type I populations. Ann. Math. Statist., 35, 1815-1818.

MV Pareto type I and MVN. Distributions of (i) order statistics in sample of size n, (ii) ranges in samples of sizes 2 and 3 for MVN, (iii) ranges in Pareto for n = 2, 3.

749. Mardia, K.V. (1964): Exact distributions of extremes, ranges and midranges in samples from any multivariate population. J. Indian Statist. Assoc., 2, 126-130.

General MV distributions. Expressions for the joint distributions of the (i) maxima, minima (ii) ranges (iii) midranges. Special cases of n = 2.

750. Mardia, K.V. (1967): Correlation of the ranges of correlated samples. Biometrika, 54, 529-539.

BV distributions. Joint p.d.f. of the ranges of X and Y. Correction of Hartley's results. Existence of the correlation of the ranges. Expansion of the correlation coefficient in powers of ρ^2. BVN case. Tables.

751. Mardia, K.V. (1970): Measures of multivariate skewness and kurtosis with applications. <u>Biometrika</u>, <u>57</u>, 519-530.

MV populations and extensions of notions of skewness and kurtosis. General distribution. Asymptotic distributions of the measures of skewness and kurtosis in MVN case. Tests for multivariate skewness and kurtosis being zero. Tests for multivariate normality. Applications to some data. Effects of non-normality on Hotelling's one sample T^2-test.

752. Mardia, K.V. (1971): The effect of nonnormality on some multivariate tests and robustness to nonnormality in the linear model. <u>Biometrika</u>, <u>58</u>, 105-121.

Box and Watson procedures applied to MV case. Permutation moments of Pillai's V statistic under assymption of symmetry of the error distribution. Expression in terms of the normal moments. Unconditional moments. Approximation to the permutation distribution by beta distribution. MANOVA in a special case and robustness of some tests: permutation moments of V in the special case. Randomization tests for MANOVA. Robustness of Mahalanobis' D^2 with equal and unequal sample sizes. Effects of nonnormality on tests for equality of covariances. Accuracy of the beta approximation when sampling from some nonnormal BV populations. Comparison of α.

753. Marriott, F.H.C. (1971): Practical problems in a method of cluster analysis. <u>Biometrics</u>, <u>27</u>, 501-514.

Classification based on generalized (within-group) variance. Choice of number of groups. A study of effect of heterogenity of dispersion matrices, varying proportions of mixing distributions, grouping of data, linear dependence, concomittant observations. Sampling distribution of criterion.

754. Marsaglia, G. (1957): A note on the construction of a multivariate normal sample. <u>IRE Trans. Information Theory</u>, <u>3</u>, 149.

A method for generating a MVN sample of a predetermined covariance structure. Start with N(0, 1) variables and perform linear transformations. Discussion of the linear transformation and its construction.

755. Marsaglia, G. (1963): Expressing the normal distribution with covariance matrix A+B in terms of one with covariance matrix A. <u>Biometrika</u>, <u>50</u>, 535-538.

If X is MVN(0,A) and F(A,α) = P{X < α}, then F[A+B,α] = E[F(A,α-η)], where η is MVN(0,B). Application of the result to establish a reduction formula for determining probability contents in positive quadrants and half infinite intervals.

756. Marsaglia, G. (1964): Conditional means and covariances of normal variables with singular covariance matrix. <u>J</u>. <u>Amer</u>. <u>Statist</u>. <u>Assoc</u>., <u>59</u>, 1203-1204.

(Y, Z) are MVN with zero mean and covariance Σ, not necessarily non-singular. Expressions for conditional mean and variances of Y given Z = z, in terms of pseudo-inverses of Σ_{ZZ}.

757. Marshall, A.W. and Olkin, I, (1967): A multivarate exponential distribution. <u>J</u>. <u>Amer</u>. <u>Statist</u>. <u>Assoc</u>., <u>62</u>, 30-44.

BV exponential distribution and its derivation. Comparison of various types of such distributions. Some properties. MV generalizations.

758. Marshall, A.W. and Olkin, I, (1968): A general approach to some screening and classificiation problems. <u>J</u>. <u>Roy</u>. <u>Statist</u>. <u>Soc</u>., <u>B</u>, <u>30</u>, 407-443.

Classification on the basis of a variable Y which is costly, inconvenient or impossible to observe. Observe $X = (x_1, x_2, \ldots, x_k)$ *which is correlated with Y. Conditional distribution of Y|X assumed known. Optimality judged in terms of a given loss function. Minimum risk procedures. Constant, quadratic and exponential loss when conditional distribution is normal. Classification when some variables are discrete. Observation of X sequentially. Examples: measuring instrument errors, calibration, screening for presence or absence of condition or disease. Discussion by several statisticians.*

759. Martin, D.C. and Bradley, R.A. (1972): Probability models, estimation, classification for multivariate dichotomous populations. <u>Biometrics</u>, <u>28</u>, 203-221.

Classification procedures using multivariate dichotomous variables. Development of probability model involving orthogonal polynomials. ML estimation. Classification based on a minimum risk decision rule. Examples.

Massey, F.J. (1965): See Dunn, O.J. and Massey, F.J., #268.

760. Massy, W.F. (1965): Principal components regression in exploratory statistical research. J. Amer. Statist. Assoc., 60, 234-256.

Review of principal components analysis. Use of principal components as dependent variables in regression analysis. Standardized coefficients of regression. Comparison with results obtained by classical multiple regression analysis. Examples. Interpretations.

761. Mathai, A.M. (1966): On multivariate exponential type distributions. J. Indian Statist. Assoc., 4, 143-154.

Characterization of multivariate distributions by conditional distribution of independent random variables and their sum being of exponential type. Examples of interest: multivariate exponential and multivariate normal.

762. Mathai, A.M. and Saxena, R.K. (1969): Distribution of a product and the structural set-up of densities. Ann. Math. Statist., 40, 1439-1448.

H-function and its properties. Expansion of H-functions. Distribution of product using Mellin transforms and inverse Mellin transforms. Special cases: product of likelihood ratios, multiple correlation coefficient, products of independent noncentral Beta variates, products of independent simple correlation coefficients.

763. Mathai, A.M. (1970): Statistical theory of distributions and Meijer's G-functions. Metron, 28, 122-146.

Meijer's G-function. Contour integration and theory of residues for series expansion. Evaluation of percentage points for distributions. Inverse Mellin transforms. Illustration of techniques: (i) ratios of Wishart determinants, (ii) testing hypotheses on regression coefficients, (iii) sphericity test.

764. Mathai, A.M. and Rathie, P.N. (1970): The exact distribution of Votaw's criteria. Ann. Inst. Statist. Math., 22, 89-116.

Votaw's criteria of bipolarity. LR under the null. Meijer G and Fox's H functions and applications. Mellin transform method. Most general null distribution in series using residue theorem and Riemann zeta function.

765. Mathai, A.M. (1971): On the distribution of the likelihood ratio criterion for testing linear hypotheses on regression coefficients. Ann. Inst. Statist. Math., 23, 181-197.

Mellin transform of the LR statistic and its inversion in the null case. General case by use of theory of residue and psi and zeta functions. Tables of percentage points.

766. Mathai, A.M. (1971): An expansion of Meijer's G-function and the distribution of products of independent beta variates. South African Statist. J., 5, 71-90.

Meijer's G- and Fox's H- functions and their expansions. Application to the distribution of the product of beta variates, Wilks' LR criteria, Votaw's criterion, sphericity test, all under null.

767. Mathai, A.M. and Rathie, P.N. (1971): The exact distribution of Wilks' criterion. Ann. Math. Statist., 42, 1010-1019.

Wilks' LR criterion for MV linear hypothesis. Test of $H_0: \beta_1 = B$. *Mellin transform and its inversion. General case.*

768. Mathai, A.M. and Rathie, P.N. (1971): The exact distribution of Wilks' generalized variance in the noncentral linear case. Sankhya, A, 33, 45-60.

Generalized variance in the noncentral linear case. The Mellin transform and its inversion by G-functions. Expansion by calculus of residues, ψ *and zeta functions.*

769. Mathai, A.M. and Rathie, P.N. (1971): The problem of testing independence. Statistica, 31, 673-688.

The LR criterion for $H_0: \Sigma_{ij} = 0$ $(i \neq j)$, *in the null case. Meijer's G-function by inverting the Mellin transforms. Expansions.*

770. Mathai, A.M. (1972): The exact non-central distribution of the generalized variance. Ann. Inst. Statist. Math., 24, 53-65.

Exact non-central distribution of Wilks' generalized variance in terms of computable functions involving zonal polynomials, psi and zeta functions. Inverse Mellin transforms. Meijer's G-function and its series representation. Calculus of residues.

771. Mathai, A.M. (1972): The exact distributions of three multivariate statistics associated with Wilks' concept of generalized variance. Sankhya, A, 34, 161-170.

Exact distribution of (i) ratio of two independent sample generalized variances, (ii) sample generalized variance, (iii) ratio of the sample generalized variance to its minors. Mellin transforms which are products of gamma functions. Properties of Meijer's G-function, ψ-function and ζ-function. Computation of percentage points.

Mathai, A.M. (1972): See Gordon, F.S. and Mathai, A.M., #392.

Mattson, R.L. (1966): See Peterson, D.W. and Mattson, R.L., #874.

772. Matusita, K. (1966): A distance and related statistics in multivariate analysis. Multivariate Anal.-I, [Proc. 1st Internat. Symp., Krishnaiah, ed.], 187-200.

Definition of "distance" between populations. Applications to MVN. Measure of distance or affinity between populations and between population and sample. Distributions of measures. Similarity with generalized T^2 distribution.

773. Matusita, K. (1967): On the notion of affinity of several distributions and some of its applications. Ann. Inst. Statist. Math., 19, 181-192.

Extension of the measure of affinity to k populations. Properties of measure. Tests of hypotheses based on "affinity" measure: (i) equality of k mean vectors, (ii) equality of k covariance matrices. Test statistics are functions of LR test statistic.

774. Matusita, K. (1967): Classification based on distance in multivariate Gaussian cases. Proc. Fifth Berkeley Symp. Math. Statist. Prob., 1, 299-304.

Classification rules based on a distance function (see #772). Application to MVN distribution. Evaluation of error rates. Classification using linear function of observation vectors.

Maurice, R. (1967): See Freeman, H., Kuzmack, A., and Maurice, R., #312.

Maxwell, A.E. (1960): See Birnbaum, A. and Maxwell, A.E., #111.

775. Mayer, L.S. (1971): A method of cluster analysis when there exist multiple indicators of a theoretic concept. Biometrics, 27, 143-155.

Partitioning a finite set of observations into one of two groups. Adjusted Mahalanobis' distance function and an algorithm based on it. Examples.

McDonald, L.L. (1969): See Srivastava, J.N. and McDonald, L.L., #1069.

McDonald, L.L. (1970): See Srivastava, J.N. and McDonald, L.L., #1070.

776. McDonald, L.L. (1971): On the estimation of missing data in the multivariate linear model. Biometrics, 27, 535-543.

Estimation of multiresponse missing data by the minimization of the trace of sum of squares and products matrix for error. Equivalence to a stepwise univariate procedure. BLUE and MANOVA with the estimated missing data. Estimability and testing of hypotheses of the type H_0: $C\xi = 0$. An example due to Immer.

777. McDonald, L.L. and Milliken, G.A. (1971): Multivariate tests for nonadditivity: a general procedure. Tech. Rep. #20, Dept. Statist., Kansas State U.

Extension of Milliken-Graybill work to multiresponse set up. Test of the additivity model. Roy's Union-Intersection principle and test based on the largest root of $S_h S_e^{-1}$. Distribution of the largest root under the null hypothesis.

McDonald, L.L. (1971): See Srivastava, J.N. and McDonald, L.L., #1071.

McDonald, L.L. (1971): See Srivastava, J.N. and McDonald, L.L., #1072.

778. McDonald, L.L. (1972): A multivariate extension of Tukey's one degree of freedom for nonadditivity. J. Amer. Statist. Assoc., 67, 674-675.

Particular model under #777. A two-way cross-classification is the model in this case. The test is the same as above, namely Roy's largest root criterion.

779. McDonald, L.L. (1972): Tests for the general linear hypothesis under the multiple design multivariate linear model. Tech. Rep. #26, Dept. Statist., Kansas State U.

Definition of multiple design, multivariate linear model. Generalization of the standard MANOVA model to permit possibility of different 'design' matrices. Use of Union-Intersection principle for testing the general linear hypothesis. Maximum root criterion. Confidence regions for location parameters and their functions. Application.

780. McDonald, R.P. (1968): A unified treatment of the weighting problem. Psychometrika, 33, 351-381.

General procedure for obtaining weighted linear combinations of variables. Invariance properties with respect to transformation of variables. Special cases of procedure: multiple regression, canonical correlation, principal component analysis, factor analysis.

781. McFadden, J.A. (1960): Two expansions for the quadrivariate normal integral. Biometrika, 47, 325-333.

Evaluation of $Pr(X > 0)$ when X is $N(0, \Sigma)$. Series solution for probability when the correlation matrix has a specified structure: (i) Markovian, (ii) zero elements except on diagonal and adjacent to it. Two cases related by transformation. Evaluation of fourth product moment.

782. McKeon, J.J. (1965): Canonical analysis: some relations between canonical correlation, factor analysis, discriminant function analysis, and scaling theory. Psychomet. Monog., 13, 1-43.

General structure of canonical analysis. Unified treatment using pseudo-variates and conditional inverses. Special cases: canonical correlation analysis, factor analysis, discriminant analysis. Data based on categorical responses. Tests of hypotheses: independence, partial independence, rank. Tests show interrelationships among techniques. Examples.

783. McLachlan, G.J. (1972): An asymptotic expansion for the variance of the errors of misclassification of the linear discriminant function. Austral. J. Statist., 14, 68-72.

Probabilities of misclassification with Anderson's linear classificatory statistic. Application of a theorem of Cramer to determine the expansions and the variances of the probabilities of misclassification to $O(n^{-3})$. Example.

784. McMahon, E.L. (1964): An extension of Price's theorem. IEEE Trans. Information Theory, 10, 168.

(X, Y) is BVN. Partial differential equation satisfied by E{f(x, y)} for any function f(x, y) of X and Y. Extension of Price's theorem.

McNee, R.C. (1960): See Danford, M.B., Hughes, H.M. and McNee, R.C., #211.

785. McNeil, D.R. (1968): The asymptotic powers of multivariate tests with grouped data. J. Roy. Statist. Soc., B, 30, 338-348.

Investigation of the effect of grouping on the performance of Hotelling's T^2. Asymptotic distribution of the statistic obtained from grouped data. Expression for the asymptotic relative efficiency. Grouping effected a signed observations or digitized observations. Special cases. Asymptotic power of Bennett's sign test.

786. McNolty, F. and Tomsky, J. (1972): Some properties of special-functions, bivariate distributions. Sankhya, B, 34, 251-264.

Special function distributions, i.e., distributions having special functions as the densities. Univariate cases discussed in general. Multivariate extension of modified Bessel function (type II) distribution. Generalized Wishart-Poisson distribution. Characteristic function. Partitioned matrix. Regenerative property.

787. Mehta, J.S. and Gurland, J. (1969): Some properties and an application of a statistic arising in testing correlation. Ann. Math. Statist., 40, 1736-1745.

LR test for $\rho = \rho_0$ when $\sigma_1^2 = \sigma_2^2 = \sigma^2$ in BVN. Non-null distribution of LR test statistic u when $\sigma_1^2 \neq \sigma_2^2$ and $\rho \neq \rho_0$. First two moments of u. Use of statistic u in problem of estimating differences in means $\mu_1 - \mu_2$ in BVN when observations are missing for one variable. Example.

788. Mehta, J.S. and Gurland, J. (1969): Testing equality of means in the presence of correlation. Biometrika, 56, 119-126.

Test of $\mu_1 = \mu_2$ when X is BVN (hypothesis of axial symmetry). Preliminary test of significance for $\rho = 0$. Testing equality of means when some observations from one variable are missing. Size and power of tests examined. Examples.

789. Mehta, M.L. and Gaudin, M. (1960): On the density of eigenvalues of a random matrix. Nuclear Phys., 18, 420-427. [Correction: 22 (1961), 340].

The integration of the joint density of the latent roots of a random matrix to determin the marginal of a single root. Expression in terms of Hermite polynomials. Relationship with the level density of heavy nuclei.

790. Melton, R.S. (1963): Some remarks on failure to meet assumptions in discriminant analyses. Psychometrika, 28, 49-53.

Construction of linear discriminant function when the mean and variance are known. Case of $\Sigma_1 \neq \Sigma_2$.

791. Memon, A.Z. and Okamoto, M. (1970): The classification statistic W* in covariate discriminant analysis. Ann. Math. Statist., 41, 1491-1499.

Classificatory procedures in the presence of covariates. Generalization of Anderson's W statistic and of Okamoto's results concerning the same, to the case when W is adjusted for the covariates. Asymptotic distribution of the standardized statistic W. Probabilities of misclassification. The comparison of three procedures based on x alone, the whole vector (x, y) or x corrected for y. Asumptotic relative efficiency.*

792. Memon, A.Z. and Okamoto, M. (1971): Asymptotic expansion of the distribution of the Z statistic in discriminant analysis. J. Multivariate Anal., 1, 294-307.

John's discriminant function Z when Σ *is assumed equal but unknown. Some properties of Z and its standard form. Asymptotic standardized distribution to terms of order* O_3. *Probabilities of misclassification and their total value. Comparison with Anderson W statistic. Table of coefficients of the linear and quadratic terms.*

Mickey, M.R. (1968): See Lachenbruch, P.A. and Mickey, M.R., #701.

Mickey, M.R. (1972): See Jenden, D.J., Fairchild, M.D., Mickey, M.R., Silverman, R.W. and Yale, C., #513.

793. Mickey, R. (1959): Some bounds on the distribution functions of the largest and smallest roots of normal determinantal equations. Ann. Math. Statist., 30, 242-243.

Roots of the determinantal equation $|S_1-\lambda S_2| = 0$ *when* $\Sigma = I$, *or equal in both populations: Minimum and maximum roots. Inequalities and bounds for the distribution functions of these roots in terms of* $G(x)$, *the distribution function of F-distribution.*

Mielke, P.W., Jr. (1972): See Wu, S., Williams, J.S. and Mielke, P.W. Jr., #1180

Mijares, T.A. (1959): See Pillai, K.C.S. and Mijares, T.A., #877.

794. Mijares, T.A. (1961): The moments of elementary symmetric functions of the roots of a matrix in multivariate analysis. Ann. Math. Statist., 32, 1152-1160.

Properties of completely homogeneous symmetric functions. Moments of Pillai's $V_1^{(s)}$. *Properties of compound matrices. Moments in general of elementary symmetric functions of the roots. Moments of* $V_j^{(s)}$.

795. Mijares, T.A. (1964): On elementary symmetric functions of the roots of a multivariate matrix: distributions. Ann. Math. Statist., 35, 1186-1198.

Joint distribution of elementary symmetric functions of the roots of $|T_1-\theta(T_1+T_2)| = 0$. *Moments and product moments expressed in determinantal form.*

796. Mikhail, N.N. (1965): A comparison of tests of the Wilks-Lawley hypothesis in multivariate analysis. Biometrika, 52, 149-156.

Comparison of MANOVA tests: (i) Wilks' LR test, (ii) Lawley's, (iii) Pillai's. Power comparisons. Approximation of noncentral distributions. Equating moments of distributions: Patnaik's method of approximating distributions. Calculations are restricted to bivariate case.

797. Mikhail, N.N. (1967): The generalization of Penrose's criterion (D-statistic) in the p-variate case. Egyptian Statist. J., 11, 21-36.

Extension of Penrose's criterion to case of p-dimensions. Expectation under null hypothesis. Variance for $p = 2, 3$. *Application of polykays. Generalized k-statistics. Randomization bases of multivariate tests.*

Mikhail, W.F. (1961): See Roy, S.N. and Mikhail, W.F., #976.

798. Mikhail, W.F. (1962): On a property of a test for the equality of two normal dispersion matrices against one-sided alternatives. Ann. Math. Statist., 33, 1463-1465.

Testing H_0: $\Sigma_1 = \Sigma_2$ using largest or smallest characteristic root of $S_1S_2^{-1}$. Considers alternatives: (i) $\gamma(min) > 1$, (ii) $\gamma(max) < 1$, (iii) $\gamma(max) > 1$, (iv) $\gamma(min) < 1$, where γ's are characteristic roots of $\Sigma_1\Sigma_2^{-1}$. Test procedures suggested by Roy and Gnanadesikan. Monotonicity characteristic of power function.

Mikhail, W.F. (1970): See Roy, S.N. and Mikhail, W.F., #980.

799. Miller, K.S. (1964): Distributions involving norms of correlated Gaussian vectors. Quart. Appl. Math., 22, 235-243.

Rayleigh random variable: norm of a MVN vector. Distributions of products and ratios. Modified Bessel functions. Moments of products and ratios. Hypergeometric functions.

800. Miller, K.S. (1965): Some multivariate density functions of products of Gaussian variates. Biometrika, 52, 645-646.

Two theorems proved: (i) X is MVN with mean zero and inverse variance matrix Σ, Y is MVN with mean zero and inverse variance matrix wI then the joint p.d.f. of $(z_1, z_2, \ldots, z_n)$ is found when $z_i = x_i|Y|$ (ii) U is MVN and partitioned as (x, y). Then joint p.d.f. of $(z_1, \ldots, z_p)$, where $z_j = x_j y$, is found. Here |Y| indicates norm of Y.

801. Miller, K.S. (1968): Some multivariate t-distributions. Ann. Math. Statist., 39, 1605-1609.

Several distributions are derived from the MVN. In most cases the random variables are of the form $x_i/|Y|$, $x_i/|Z|$, where X, Y, Z are all MVN. Here |Y| denotes the norm of Y. Ratios of normal variables which arise from the same MVN distributions.

802. Miller, R.G. (1962): Statistical prediction by discriminant analysis. Meteorol. Monog., 4, (25), 1-54.

A survey and review article detailing the multiple discriminant analysis and the test procedures related to it. Applications to meteorological situations.

Milliken, G. (1969): See Graybill, F.A. and Milliken, G., #398.

Milliken, G.A. (1971): See McDonald, L.L. and Milliken, G.A., #777.

Milton, R. (1970): See Gurland, J. and Milton, R., #427.

803. Mitra, S.K. (1969): Some characteristic and noncharacteristic properties of the Wishart distribution. Sankhya, A, 31, 19-22.

If S is Wishart, the following properties are true: (i) $a'\Sigma a \neq 0$, then $a'Sa/a'\Sigma a$ is χ^2, (ii) $a'\Sigma a = 0$ implies $a'Sa = 0$ with probability one. The converse is not true unless rank of Σ is 1. That is, these two properties do not characterize Wishartness of S. Some distributions other than Wishart possess these properties. Characterization of the Wishart distribution on the basis of these two conditions in addition to a minimal set of others. Singular Wishart and its characteristic function.

804. Mitra, S.K. (1970): A density-free approach to the matrix variate beta distribution. Sankhya, A, 32, 81-88.

Definition of a MV beta matrix in terms of the Wishart matrix. Derivation of the properties by density-free approach by applying Rao's results with respect to the Wishart distribution. Submatrices have MV beta distribution and a submatrix is independently distributed of a conditional matrix. Independence of the ratios of successive determinants (Rao's theorem). Transformations by orthogonal matrices. Non-characteristic property of MV beta.

805. Mitra, S.K. (1970): Analogues of multivariate beta (Dirichlet) distributions. Sankhya, A, 32, 189-192.

Multivariate Dirichlet distribution and a density-free approach to its properties. Relationship with Wishart and MV beta exploited. Density function given. Product of a univariate beta and a MV Dirichlet is a new MV distribution. Density function. Distribution of the 'product' of a MV beta and Dirichlet variates. Relationship to Mauchly's sphericity criterion.

Money, A.H. (1972): See Troskie, C.G. and Money, A.H., #1139.

Moore, A.W. (1968): See Horton, I.F., Russell, J.S. and Moore, A.W., #478.

806. Moran, P.A.P. (1967): Testing for correlation between non-negative variates. Biometrika, 54, 385-394.

Bivariate gamma and negative exponential distributions. Properties of the BV gamma distributions. Tests for the correlation begin zero in BV negative exponential. Approximations. Moments of the statistic.

807. Moran, P.A.P. (1969): Statistical inference with bivariate gamma distributions. Biometrika, 56, 627-634.

Generating the BV gamma from the BVN distribution. Estimation of the parameters by ML and some tests of significance. Applications to crossover designs.

808. Morgenthaler, G.W. (1961): Some circular coverage problems. Biometrika, 48, 313-324.

Approximate estimation of the average area of a circle centered at (0,0) when it is covered by (i) a random circle with a BVN distribution, (ii) n random circles whose centers are independent BVN, (iii) a cluster of circles with the center of the cluster being BVN.

Morishima, H. (1969): See Hashiguchi, S. and Morishima, H., #450.

Morris, C. (1972): See Efron, B. and Morris, C., #288.

Morris, C. (1972): See Efron, B. and Morris, C., #289.

Morris, C. (1972): See Sclove, S.L., Morris, C. and Radhakrishnan, R., #1013.

Morris, R.H. (1957): See Jackson, J.E. and Morris, R.H., #496.

Morrison, D.F. (1961): See Birren, J.E. and Morrison, D.F., #112.

809. Morrison, D.F. (1962): On the distribution of the sums of squares and cross products of normal variates in the presence of intra-class correlation. Ann. Math. Statist., 33, 1461-1463.

The sample variance matrix and its distribution when the population matrix is of a particular form. Partitioned submatrices are of the intra-class correlation pattern.

810. Morrison, D.F. (1970): The optimal spacing of repeated measurements. Biometrics, 26, 281-290.

Repeated measurements experiments. Time ordering. Expected values and covariance structure of responses as functions of position on scale. Spacing of treatments so as to maximize power of T^2 test for equal treatment effects and equality of mean vectors in two sample case. Covariance structure: (i) Wiener, (ii) Markov. Alternative hypothesis: mean vectors whose elements are linear or quadratic functions of scale parameter.

811. Morrison, D.F. (1971): The distribution of linear functions of independent F variates. J. Amer. Statist. Assoc., 66, 383-385.

Distribution of $\ell_1F_1+\ell_2F_2$ where F_1, F_2 are independent F variates. Exact distribution: series of incomplete beta functions. Approximation to F distribution by equating first two cumulants. Application to testing $\mu = \mu_0$ under an assumption of compound symmetry.

812. Morrison, D.F. (1971): Expectations and variances of maximum likelihood estimates of the multivariate normal distribution parameters with missing data. J. Amer. Statist. Assoc., 66, 602-604.

Exact expectations and variances of MLE of μ and Σ for a single incomplete and of complete variates. Biases, variances and MSE of estimators. Comparison with ML estimates based on complete observation vectors. Efficiency is a function of multiple correlation between missing variate and complete vector.

813. Morrison, D.F. (1972): The analysis of a single sample of repeated measurements. Biometrics, 28, 55-71.

Use of multivariate techniques for analysis of repeated measurements experiments. Equality of means with general covariance matrix and with special covariance patterns: (i) symmetric covariance matrix, (ii) reducible covariance matrices, (iii) compound symmetry, (iv) covariance matrix known up to a constant. Comparison of methods in terms of average squared lengths of their simultaneous confidence intervals. LR tests. Incomplete data. Test for paired t situation with missing values on one variate and known correlation. ML estimates with incomplete data. Confidence intervals. Comparison of missing value method with paired t based on only complete vectors.

Mosimann, J.E. (1960): See Jolicoeur, P. and Mosimann, J.E., #539.

814. Mosimann, J.E. (1970): Size allometry: Size and shape variables with characterizations of the lognormal and generalized Gamma distributions. J. Amer. Statist. Assoc., 65, 930-945.

Study of differences in shape as associated with size: allometry. Size variables and shape vectors. Isometry: independence of size and shape. Jolicoeur's test for isometry. Characterizations: lognormal, gamma and generalized gamma. Contrast with linear functional relationship. Size and shape interpretation using principal components or discriminant analysis.

815. Mostafa, M.D. and Mahmoud, M.W. (1964): On the problem of estimation of the bivariate lognormal distribution. Biometrika, 51, 522-527.

Bivariate lognormal distribution and its relationships with BVN. Estimation of regressions: mean, median, modal. Mean and variance of estimators. Comparison of estimators.

816. Muddapur, M.V. (1968): On the power of the test for multiple correlation coefficient. Metron, 27, 214-219.

Asymptotic noncentral (unconditional) distribution of $(N-p)R^2/(p-1)(1-R^2)$: noncentral F with $(p-1)$ and $(N-p)$ degrees of freedom and noncentrality parameter $(N-1)\bar{R}^2/2$. Charts for calculating power of test.

Mudholkar, G.S. (1964): See Das Gupta, S., Anderson, T.W. and Mudholkar, G.S., #217.

817. Mudholkar, G.S. (1965): A class of tests with monotone power functions for two problems in multivariate statistical analysis. Ann. Math. Statist., 36, 1794-1801.

Relationship between MANOVA tests and tests for independence. Properties of invariant tests and power. Symmetric gauge functions and their properties. Convex functions of matrices. Some inequalities on the symmetric gauge functions of the roots. Construction of tests with monotone power, using symmetric gauge functions. MANOVA and independence.

818. Mudholkar, G.S. (1966): On confidence bounds associated with multivariate analysis of variance and non-independence between two sets of variates. Ann. Math. Statist., 37, 1736-1746.

A detailed discussion of Roy's Union-Intersection principle and simultaneous confidence bounds. Application to MANOVA and independence between sets of variates. Symmetric gauge functions and their properties. Some inequalities on symmetric gauge functions of characteristic roots. Simultaneous confidence bounds on symmetric gauge functions of the roots of a matrix in the independence case. MANOVA dealt with from the independence case.

819. Muirhead, R.J. (1970): Systems of partial differential equations for hypergeometric functions of matrix argument. Ann. Math. Statist., 41, 991-1001.

Hypergeometric function ${}_2F_1$ of matrix argument. Partial differential equations. Zonal polynomials. Expansions and differentiations. Differential equations for ${}_1F_1$ and ${}_0F_1$.

820. Muirhead, R.J. (1970): Asymptotic distributions of some multivariate tests. Ann. Math. Statist., 41, 1002-1010.

Applications of the partial differential equations for ${}_2F_1$ and related functions. General theory of asymptotic distributions derived from the partial differential equations. Evaluation of the coefficients. Examples: Hotelling's T_0^2, Pillai's $V^{(m)}$, largest root of a covariance matrix.

821. Muirhead, R.J. (1972): On the test of independence between two sets of variates. Ann. Math. Statist., 43, 1491-1497.

Application of #820 to derive the asymptotic non-null distribution for the modified LR criterion. Coefficients of expansion to $O(n^{-3})$. Numerical comparisons.

822. Muirhead, R.J. (1972): The asymptotic noncentral distribution of Hotelling's generalized T_0^2. Ann. Math. Statist., 43, 1671-1677.

Partial differential equation satisfied by the Bessel function of the second kind, with matrix argument. Moment generating function of Hotelling's T_0^2 and its expansion. Evaluation of the coefficients. Inversion of the MGF. Asymptotic non-null distribution.

Mukherjee, G.D. (1964): See Roy, S.G. and Mukherjee, G.D., #964.

823. Mukherjee, R. and Bandyopadhyay, S. (1964): Social research and Mahalanobis' D^2. Contrib. Statist., (Mahalanobis Volume, Rao, ed.), 259-282.

Examples of Mahalanobis' D^2. Stepwise D^2 analysis. Appropriate tests of significance. Interpretation of results.

824. Mukherji, V. (1967): "A note on maximum likelihood" - a generalization. Sankhya, A, 29, 105-106.

Generalization of Watson's result for a two-variate model. Instrument variables. Equating Σ with the sample value S. Estimate asymptotically equivalent to MLE of Σ.

Murthy, V.K. (1960): See Roy, J. and Murthy, V.K., #962.

825. Musket, S.F. (1971): An evaluation of Kendall's order statistic method of discriminant analysis and related studies. Tech. Rep. #112, Dept. Statist., Southern Methodist U.

Kendall's method of discriminant analysis using order statistics. Effects of unequal covariance matrices. Bartlett-Please model. Monte Carlo studies. Probabilities of misclassification for linear discriminant functions. Effect of nonnormality: Cauchy, uniform, lognormal distributions (independent). Modification of Bartlett-Please model. Comparison of Kendall's method with Feldman, Klein, Honigfeld method. Probabilities of misclassification.

826. Mustafi, C.K. (1968): On the proportion of observations above sample means in a bivariate normal distribution. Ann. Math. Statist., 39, 1350-1353.

Joint distribution of proportion of observations above $\bar{X}$ and proportion above $\bar{Y}$ when X and Y are BVN. Asymptotic distribution is BVN. Order statistics in BVN. Generalization to proportions above and below mean.

827. Nabeya, S. (1961): Absolute and imcomplete moments of the multivariate normal distribution. Biometrika, 48, 77-84.

Absolute moments of the type $E\{|x_1x_2x_3x_4|\}$. Exact expression involving $E\{sgn(x_1x_2x_3x_4)\}^2$. Evaluation of $E\{|x_1^{n_1}x_2^{n_2}x_3^{n_3}x_4^{n_4}|\}$ for $2 = n_1 \geq n_2 \geq n_3 \geq n_4 \geq 1$. Fourth moment of mean deviation and of Gini's mean difference on samples from MVN.

828. Nadler, J. (1967): Bivariate samples with missing values. Technometrics, 9, 679-682.

Answer to a query concerning missing values in BVN. Estimation of Σ when number of observations on X and Y are not equal.

Nagao, H. (1968): See Sugiura, N. and Nagao, H., #1104.

Nagao, H. (1971): See Sugiura, N. and Nagao, H., #1108.

Nagarsenker, B.N. (1971): See Pillai, K.C.S. and Nagarsenker, B.N., #901.

829. Nagarsenker, B.N. and Pillai, K.C.S. (1972): The distribution of the sphericity test criterion. ARL 72-0154, Wright Patterson AFB, Ohio.

Test of hypothesis H_0: $\Sigma = \sigma^2 I$. Sphericity criterion and its null distribution by use of three methods: (i) Mellin transforms, (ii) incomplete Beta integrals, (iii) Gamma series. Approximations by Wilks-Tukey's method, Box's expansion of the characteristic functions, and Mauchly's fitting of Pearson curve. Comparison of the approximations with Davis' inversion of Box's series by Cornish-Fisher expansion. Tables of comparison of approximate percentage points. Tables of the exact percentage points.

Nagarsenker, B.N. (1972): See Pillai, K.C.S. and Nagarsenker, B.N., #905.

830. Nakajima, N. and Tsii, K. (1967): Multidimensional tolerance regions based on a large sample. Ann. Inst. Statist. Math., 19, 439-449.

A general theory of multidimensional tolerance regions of confidence level β (asymptotic) with content γ. Use of Stokes' theorem. Ellipsoidal regions for MVN and a verification of Wald's results.

831. Nandi, H.K. (1963): On the admissibility of a class of tests. Calcutta Statist. Assoc. Bull., 12, 13-18.

An extension of Stein's theorem on admissibility. Admissibility of Roy's Union-Intersection tests. Applications to Hotelling's T^2 and tests for equality of dispersion matrices.

832. Nandi, H.K. (1965): On some properties of Roy's Union-Intersection tests. Calcutta Statist. Assoc. Bull., 14, 9-13.

Optimality of Roy's Union-Intersection tests: consistency, unbiasedness and admissibility based on the properties of the components. Simultaneous confidence sets for the hypotheses.

833. Nath, G.B. (1971): Estimation in truncated bivariate normal distributions. Appl. Statist., 20, 313-319.

MLE in truncated BVN: (i) singly or doubly truncated, (ii) linearly truncated. Expressions for the asymptotic variances (information limit). Numerical example.

834. Nathanson, J.A. (1971): An application of multivariate analysis in astronomy. Appl. Statist., 20, 239-249.

A discriminant analysis of astronomical data.

Naylor, J.C. (1965): See Wherry, R.J., Naylor, J.C., Wherry, R.J. Jr., and Fallis, R.F., #1165.

Neave, H.R. (1971): See Bhattacharyya, G.K., Johnson, R.A. and Neave, H.R., #106.

835. Nel, D.G. (1971): The hth moment of the trace of a noncentral Wishart matrix. South African Statist. J., 5, 41-52.

$E\{tr\ A\}^h$ when A is Wishart $[\sigma^2 I,\ n,\ \Omega]$ for $h = 1, 2, 3, 4$. Series expansion. Elementary symmetric functions. General expression by induction. Tables of coefficients.

836. Nel, D.G. (1972): On the moments of the trace of noncentral Wishart matrices and submatrices. South African Statist. J., 6, 93-102.

Elementary symmetric functions and differentiation with respect to a parameter. Moment generating function of a noncentral Wishart matrix. Moments of trace of noncentral Wishart matrix. Moments of traces of partitions of such a matrix..

Nelson, W.C. (1968): See Aitkin, M.A., Nelson, W.C. and Reinfurt, K.M., #18.

837. Neudecker, H. (1969): Some theorems on matrix differentiation with special reference to Kronecker matrix products. J. Amer. Statist. Assoc., 64, 953-963.

Differentiation of matrix functions: (i) Kronecker matrix products, (ii) nonKronecker (ordinary) matrix products. Several properties of the Kronecker products. Applications to Jacobians and MLE are given.

838. Nicholson, G.E. Jr. (1957): Estimation of parameters from incomplete multivariate samples. J. Amer. Statist. Assoc., 52, 523-526.

MLE when the data are incomplete. Regression techniques. Primary concern is of estimation of linear regression. Truncated distributions.

839. Nicholson, G.E. Jr. (1960): Prediction in future samples. Contrib. Prob. Statist., (Hotelling Volume, Olkin et al, eds.), 322-330.

Prediction and the use of prediction equation for a second sample. Adequacy and a measure of efficiency. Shrinkage of the multiple correlation coefficient. Distribution of the efficiency. Probability inequalities on this measure. An example.

840. Nuesch, P. (1966): On the problem of testing location in multivariate populations for restricted alternatives. Ann. Math. Statist., 37, 113-119.

Testing of the hypothesis H_0: $\mu = 0$ against H_A: $\mu > 0$. Quadratic programming techniques. LR procedure and asymptotic distribution of the LR statistic. Σ known and unknown.

841. Obenchain, R.L. (1971): Multivariate procedures invariant under linear transformations. Ann. Math. Statist., 42, 1569-1578. [Correction: 43 (1972), 1742-1743].

Procedures invariant under $L(p)$ of translations and nonsingular linear transformations. Maximal $L(p)$ invariant statistics. Geometrical property. MVN and the distribution of maximal invariants. Tests that k populations are identical. Pillai's trace criterion.

842. Obenchain, R.L. (1972): Regression optimality of principal components. Ann. Math. Statist., 43, 1317-1319.

$A_0 = \{y | \gamma_p'(y-\mu) = 0\}$, where γ_p is the vector corresponding to the smallest root of Σ and μ is the centre of the distribution. $A_1, \ldots, A_p$ be the hyperplanes corresponding to the linear regression of each variable onto the other variables. An optimal property of A_0 is that it is the best fitting a single approximation to $A_1, \ldots, A_p$. This is in the sense that $\alpha'\Sigma^{-2}\alpha \leq \lambda_p^{-2}$ and the maximum is achieved if and only if $\alpha = \gamma_p$. Optimality property of subspaces. Generalization of Okamoto's results.

843. Odell, P.L. and Feiveson, A.H. (1966): A numerical procedure to generate a sample covariance matrix. J. Amer. Statist. Assoc., 61, 199-203. [Correction: 61 (1966), 1248-1249].

A method of generating a Wishart matrix of variance matrix R and any dimensionality. Generalization of Hartley-Harris technique. Generation of the canonical form and subsequent transformation.

844. Ogawa, J. (1962): Estimation of correlation coefficient by order statistics. Contrib. Order Statist., (Sarhan and Greenberg, eds.), 283-291.

Procedures using order statistics for estimating ρ in BVN: (i) when all parameters are known, then the proportions of observations falling in six regions are used with MLE, (ii) when all the parameters unknown, then the observations are ordered along x and y axes and MLE employed with the observation proportions, (iii) variance ratios are known and $\mu = 0$, direct approach used.

845. Okamoto, M. (1961): Discrimination for variance matrices. Osaka Math. J., 13, 1-39.

Discrimination between two populations on the basis of the variance matrices when means are equal: Two cases studied (i) Σ_1, Σ_2 known. Bayes and minimax discrimination. Relationship with canonical variates. Reduction of dimensionality by use of canonical variates. Weighted χ^2 and approximations to its distribution. Probability of errors in the minimax procedure, tabled. (ii) Estimated parameters (a) when μ is known and Σ_1, Σ_2 unknown (b) all the parameters are estimated. Asymptotic distribution of eigenvalues and eigenvectors. Quadratic discriminant function and its asymptotic distribution. Reduction of dimensions. Application to a problem with twins.

846. Okamoto, M. (1963): An asymptotic expansion for the distribution of the linear discriminant function. Ann. Math. Statist., 34, 1286-1301. [Correction: 39 (1968), 1358-1359].

Anderson's classificatory statistic and its asymptotic expansion when N_1, N_2 are not equal. Expansions for probabilities of misclassification in powers of the sample size. Generalization of Bowker-Sitgreaves results. Studentization methods of Hartley and Welch.

847. Okamoto, M. and Kanazawa, M. (1968): Minimization of eigenvalues of a matrix and optimality of principal components. Ann. Math. Statist., 39, 859-863.

Some properties of functions of matrix argument. Monotonicity of such functions and their relation to characteristic roots of the matrix. Optimality property in the sense of minimization of f{E[(x-Ay)(x-Ay)']} with respect to A and y. Generalization of Rao-Darroch results.

848. Okamoto, M. (1969): Optimality of principal components. Multivariate Anal.-II, [Proc. 2nd Internat. Symp., Krishnaiah, ed.], 673-685.

Definition of principal components. Optimality properties: variation, information loss and correlation. Minimization or maximation. Unified treatment of principal components.

Okamoto, M. (1969): See Asoh, Y.H. and Okamoto, M., #52.

Okamoto, M. (1970): See Memon, A.Z. and Okamoto, M., #791.

Okamoto, M. (1971): See Memon, A.Z. and Okamoto, M., #792.

849. Oliver, P.C.D.V. (1972): Contributions to normal sampling theory. South African Statist. J., 6, 53-82.

Distributions involving the Wishart matrix. Hotelling's T^2. Generalized variances and ratios of generalized variances. Characteristic functions and Mellin transform methods. Review of recent work.

850. Olkin, I. and Pratt, J.W. (1958): Unbiased estimation of certain correlation coefficients. Ann. Math. Statist., 29, 201-211.

Unbiased estimation of correlation coefficients: (i) ρ in BVN, (ii) intraclass correlation coefficient in p-variate MVN with equal variances and covariances (iii) squared multiple correlation coefficient. Sufficient statistic. Minimum variance. Estimator obtained by inverting a Laplace transform. Tables of unbiased estimators as functions of usual estimators.

851. Olkin, I. (1959): A class of integral identities with matrix argument. Duke Math. J., 26, 207-213.

Integrals involving matrix arguments. Alternative proof of Bellman's identity. Correlation identity and its generalization. Generalized Beta integral. Matrix analogue of Liouville-Dirichlet integral. Siegel-Ingham identity. Applicable in multivariate distribution and moment problems when variables have Wishart distribution. Matrix transformations.

Olkin, I. (1959): See Bush, K.A. and Olkin, I., #130.

Olkin, I. (1961): See Bush, K.A. and Olkin, I., #131.

852. Olkin, I. and Tate, R.F. (1961): Multivariate correlation models with mixed discrete and continuous variables. Ann. Math. Statist., 32, 448-465. [Correction: 36 (1965), 343-344].

Multivariate extension of Tate's work in bivariate case. Model: $(k+1)$-variate vector x has multinomial distribution and conditional distribution of the p-variate vector y for fixed x is MVN. Consider three situations (i) $k = 1$, $p > 1$, (ii) $k > 1$, $p = 1$, (iii) $k > 1$, $p > 1$. Exact and asymptotic distributions of multiple and partial correlation coefficients. Unified approach through canonical correlations. Concept of vector correlations. Similarities with standard result in MVN situation. ML estimation. Examples.

Olkin, I. (1962): See Ghurye, S.G. and Olkin, I., #351.

853. Olkin, I. and Rubin, H. (1962): A characterization of the Wishart distribution. Ann. Math. Statist., 33, 1272-1280.

Characterization of Wishart distribution based on a multivariate generalization of Lukac's characterization of gamma distribution. Theorem: If U and V are positive definite matrices which are independently distributed and (i) $U+V = WW'$ is statistically independent of $Z = W^{-1}V(W')^{-1}$, (ii) distribution of Z is invariant under transformation $Z \to \Gamma Z \Gamma'$ with Γ orthogonal, the U and V are Wishart with same scale matrix. Proof depends on set of differential equations and their solution. Difficulty of differentiating under expectation

854. Olkin, I. and Rubin, H. (1964): Multivariate beta distributions and independence properties of the Wishart distribution. Ann. Math. Statist., 35, 261-269.

Development of multivariate beta distribution from the Wishart distribution. Distributions related to multivariate beta. Analogy between χ^2 *and Wishart distributions. Independence properties of Wishart distribution. Studentization of Wishart matrix. Square root transformations.*

Olkin, I. (1965): See Cacoullos, T. and Olkin, I., #134.

Olkin, I. (1966): See Gleser, L.J. and Olkin, I., #372.

Olkin, I. (1967): See Marshall, A.W. and Olkin, I., #757.

Olkin, I. (1968): See Marshall, A.W. and Olkin, I., #758.

Olkin, I. (1969): See Ghurye, S.G. and Olkin, I., #352.

OOlkin, I. (1969): See Gleser, L.J. and Olkin, I., #374.

Olkin, I. (1969): See Madansky, A. and Olkin, I., #738.

855. Olkin, I. and Press, S.J. (1969): Testing and estimation for a circular stationary model. Ann. Math. Statist., 40, 1358-1373.

Circular model for covarinace matrix. Circular symmetry model and spherical model. Tests for symmetries in covariance matrix. Tests of hypotheses for means. LR tests. Asymptotic distributions under null and alternative. Canonical forms and MLE under circular model.

Olkin, I. (1970): See Gleser, L.J. and Olkin, I., #375.

856. Olkin, I. and Shrikhande, S.S. (1970): An extension of Wilks' test for the equality of means. Ann. Math. Statist., 41, 683-687.

Testing equality of components of mean vector in a single sample. Covariance structure: $\Sigma = \sigma^2[(1-\rho)I+\rho e'e]$. *Assume* $\mu_1 = \mu_2 = \dots = \mu_k$ *for* $k \leq p$. *Consider regions:*
(i) $\mu_1 = \mu_2 = \dots = \mu_p$, *(ii)* $\mu_1 = \mu_2 = \dots = \mu_k = \dots = \mu_p$,
(iii) $\mu_1 = \mu_2 = \dots = \mu_k$, $-\infty < \mu_j < \infty$, $j = k+1,\dots,p$.
LR tests for (i) vs. (ii), (i) vs. (iii) and (ii) vs. (vii). Central and noncentral distribution of LR statistic.

Olkin, I. (1971): See Ghurye, S.G. and Olkin, I., #353.

Olkin, I. (1972): See Gleser, L.J. and Olkin, I., #376.

Olkin, I. (1972): See Kraft, C.H., Olkin, I. and Van Eeden, C., #638.

Olshen, R.A. (1972): See Eaton, M.L. and Olshen, R.A., #284.

857. Olson, W.H. and Uppuluri, V.R.R. (1970): Characterization of the distribution of a random matrix by rotational invariance. Sankhya, A, 32, 325-328.

Characterization of the distribution of the elements of a symmetric random matrix. Equality of distributions of matrix and its orthogonal similarity transform.

Oschinsky, L. (1958): See East, D.A. and Oschinsky, L., #279.

858. Ostroviskii, I.V. (1965): The multidimensional analogue of Yu. V. Linnik's theorem on decompositions of a convolution of Gaussian and Poisson laws. Theory Prob. Appl., 10, 673-677.

Necessary and sufficient conditions for sum of two independent multivariate vectors to be convolution of Gaussian and Poisson: each must be multivariate Gaussian and multivariate Poisson. Characteristic functions.

859. Owen, D.F. and Wiesen, J.M. (1959): A method of computing bivariate normal probabilities with an application to handling errors in testing and measuring. Bell System Tech. J., 38, 553-572.

Evaluation of volumes of bivariate normal over rectangular regions. Volumes in terms of $Pr(X > h, Y > k)$ *and a cumulative* $N(0,1)$. *Tables and charts.*

860. Owen, D.B. and Steck, G.P. (1962): Moments of order statistics from the equicorrelated multivariate normal distribution. Ann. Math. Statist., 33, 1286-1291.

MVN with equal correlations. Moments and product moments of the ith order statistic for any ρ *in terms of those for* $\rho = 0$. *Tables of* μ, σ, μ_3, μ_4 *for first order statistic for* $n = 2(1)5(5)50$ *and* $\rho = -1/(n-1)$, 0, $\frac{1}{2}$, 1 *and for* μ *and* σ *of central order statistic for* $n = 4(2)50$ *with same* ρ.

Owen, D.B. (1962): See Steck, G.P. and Owen, D.B., #1085.

861. Owen, D.B., Craswell, K.J. and Hanson, D.L. (1964): Nonparametric upper confidence bounds for Pr{Y < X} and confidence limits for Pr{Y < X} when X and Y are normal. J. Amer. Statist. Assoc., 59, 906-924.

Estimation of $P\{Y < X\}$ *when (i)* (X,Y) *are BVN with all parameters unknown or only* μ_1, μ_2 *unknown (ii)* (X,Y) *are independent,* $n_1 = n_2$, $\sigma_1 = \sigma_2$ *but unknown, (iii) paired observations,* σ_1, σ_2 *and* μ *unknown. Bound on the sample size. Use of noncentral t. Tables.*

862. Owen, D.B. (1965): A special case of a bivariate noncentral t-distribution. Biometrika, 52, 437-446.

Integrals involving univariate noncentral t distribution. Definition of special functions. BV noncentral t defined as $T_1 = (X+\delta_1)/Y$, $T_2 = (X+\delta_2)/Y$, X is univariate normal and Y is square root of χ^2/ν. Evaluation of the probabilities $P\{T_1 \leq t_1, T_2 \leq t_2\}$ with the help of the special functions. Applications to two-sided sampling plans and two-sided tolerance regions.

Owen, D.B. (1966): See Bland, R.P. and Owen, D.B., #114.

Panchapakesan, S. (1969): See Gupta, S.S. and Panchapakesan, S., #424.

863. Parikh, N.T. and Sheth. R.J. (1966): Applications of bivariate normal distribution: to a stress vs. strength problem in reliability analysis. J. Indian Statist. Assoc., 4, 105-107.

$P\{X > Y\}$ where X, Y are independent and X is $N(\mu_1, \sigma_1^2)$ and Y is $N(\mu_2, \sigma_2^2)$. Y is truncated over (y_1, y_0). Use of BVN for evaluation of the integral.

864. Park, S.H. (1969): Characteristics of some bivariate distributions with different marginal distributions. Trabajos Estadist., 20, 85-102.

BV distributions when the distribution of X is not of the same form as that of Y. Comparisons with BVN.

Pathak, P.K. (1967): See Krishnaiah, P.R. and Pathak, P.K., #652.

Pathak, P.K. (1968): See Krishnaiah, P.R. and Pathak, P.K., #654.

Patil, G.P. (1968): See Bildikar, S. and Patil, G.P., #108.

865. Patil, G.P. and Boswell, M.T. (1970): A characteristic property of the multivariate normal density function and some of its applications. Ann. Math. Statist., 41, 1970-1977.

Partial differential equations satisfied by BVN density. Price's theorem and its generalization. Mixed moments of MVN. Expression for $E(X_1, X_2, \ldots, X_{2m})$. A characterization of MVN is that $\phi(t_1, \ldots, t_n)$ (the characteristic function) satisfies $\phi'(t_1, \ldots, t_n) = -t_j t_k \phi(t_1, \ldots, t_n)$ where differentiation is with respect to σ_{jk}. Other characterizations follow. Independence and its characterization by moments.

866. Patil, S.A. and Liao, S.H. (1970): The distribution of the ratios of means to the square root of the sum of variances of a bivariate normal sample. Ann. Math. Statist., 41, 723-728. [Correction: 42 (1971), 1461].

(X, Y) are BVN with zero means, common variance σ^2 and ρ as coefficient of correlation. Distribution of (i) $m(s_1^2+s_2^2)$, (ii) (T_1, T_2), where $T_1 = \bar{X}/\sqrt{m}\, s$, $T_2 = \bar{Y}/\sqrt{m}\, s$. Here m is the sample size and s_1^2, s_2^2, s^2 are all unbiased estimates of σ^2. Distribution function of (T_1, T_2) as well as the marginal of T_1 given.

867. Pawula, R.F. (1967): A modified version of Price's theorem. IEEE Trans. Information Theory, 13, 285-288.

Partial differential equations for MVN distribution and a single ordinary differential equation. Some applications.

868. Pearce, S.C. and Holland, D.A. (1960): Some applications of multivariate methods in botany. Appl. Statist., 9, 1-7.

Principal component and factor analyses of botanical data. Interpretation of the principal components and the factors.

869. Pearson, E.S. (1969): Some comments on the accuracy of Box's approximations to the distribution of M. Biometrika, 56, 219-220.

Comparison of Box's approximation in relation to Korin's (#630) work. χ^2 and F approximations. Tables.

870. Peers, H.W. (1971): Likelihood ratio and associated test criteria. Biometrika, 58, 577-587.

Accurate approximations to the power function of LR criterion for testing a simple null against a class of component alternatives: H_0: $\theta = \theta_0$ against H_A: $\theta \neq \theta_0$. Comparison of Rao's statistics with the LR criterion in terms of power. MV Edgeworth type A series expansion for the probability density. Use of moment generating function technique. Examples. BVN case.

871. Perlman, M.D. (1969): One-sided testing problems in multivariate analysis. Ann. Math. Statist., 40, 549-567. [Correction: 42 (1971), 1777].

Tests of hypothesis H_0: $\mu \in P_1$ against $\mu \in P_2$, where P_1, P_2 are positively homogeneous sets with $P_1 \subset P_2$. LR tests are developed and the power shown to tend to 1 with distance. Exact null distribution when $P_1 \equiv \{0\}$ and $P_2 \equiv \{\mu, \mu_i \geq 0, i = 1,\ldots,p\}$. Level α similar tests. Bounds for variation of ε on the null distribution of LR statistic when $P_1 \equiv \{0\}$ and $P_2 \equiv C$, a cone. Problem of testing a one-sided hypothesis against unrestricted alternatives. Tests on subsets of mean vector.

872. Perlman, M.D. (1971): Multivariate one-sided testing problems involving Fisher's discriminant function. Sankhya, A, 33, 19-34.

LR test for the hypothesis H: $\Sigma^{-1}\eta \epsilon P_1$ against K: $\Sigma^{-1}\eta \epsilon P_2$, Σ unknown and unrestricted. Here P_1, P_2 are closed homogeneous sets. MLE. Distribution of LR statistic under H: (i) one-sided alternative when $P_1 \equiv \{0\}$ and P_2 is a cone, (ii) one-sided alternatives involving a subset of components of the vector $\Sigma^{-1}\eta$, (iii) testing $\Sigma^{-1}\eta = 0$ against $\Sigma^{-1}\eta \geq 0$, (iv) one-sided alternatives, $\Sigma^{-1}\eta \in$ a cone against unrestricted alternatives, (v) one-sided hypotheses involving a subset of the elements of $\Sigma^{-1}\eta$. Power of the test.

873. Perlman, M.D. (1972): Monotonicity of the power function of Pillai's trace test. Tech. Rep. #194, School Statist., U. Minnesota.

Acceptance region and the cut-off point of Pillai's trace criterion. Use of Das Gupta, Anderson and Mudholkar result to prove the monotonicity of Pillai's criterion in each of the noncentrality parameters. MANOVA and independence tests. Four cases of relationship between p, n and r considered.

Perlman, M.D. (1972): See Das Gupta, S. and Perlman, M.D., #225.

Perlman, M.D. (1972): See Eaton, M.L. and Perlman, M.D., #285.

874. Peterson, D.W. and Mattson, R.L. (1966): A method of finding linear discriminant functions for a class of performance criteria. IEEE Trans. Information Theory, 12, 380-387.

Pattern recognition. Bayes criterion applied to a linear discriminant function. Fisher's linear discriminant function. Comparison. Application of discriminant functions to problems in pattern recognition. Discriminatory and classificatory procedures.

Pffaffenberger, R.C. (1970): See Smith, W.B. and Pfaffenberger, R.C., #1047.

875. Phipps, M.C. (1971): An identity in Hermite polynomials. Biometrika, 58, 219-222.

Extension of Runge's identity for Hermite polynomials in terms of bivariate normal variables. Test for bivariate normality based on identity.

876. Pillai, K.C.S. and Bantegui, C.G. (1959): On the distribution of the largest of six roots of a matrix in multivariate analysis. Biometrika, 46, 237-240.

Null joint distribution of s characteristic roots. Cumulative distribution function for the largest of six characteristic roots. Approximation to distribution. Tables for upper 1% and 5% points.

877. Pillai, K.C.S. and Mijares, T.A. (1959): On the moments of the trace of a matrix and approximations to its distribution. Ann. Math. Statist., 30, 1135-1140.

Distribution of Pillai $V^{(s)}$ criterion. Moment generating function. Evaluation of first four moments, β_1 and β_2. Approximation to cumulative distribution: (i) Pearsonian curve, (ii) Pillai's approximate beta distribution.

878. Pillai, K.C.S. and Samson, P. (1959): On Hotelling's generalization of T^2. Biometrika, 46, 160-168.

Distribution of $U^{(s)}$ = sum of roots of $S_2^{-1}S_1$ where s is number of non-zero roots. Moments of $U^{(s)}$ for s = 2,...,6. Approximation to upper percentage points using: (i) Pearsonian curve, (ii) Pillai's approximation. For s = 2, table comparing approximation and exact percentage points. Percentage points for approximation for s = 2, 3, 4.

879. Pillai, K.C.S. (1964): On the moments of elementary symmetric functions of the roots of two matrices. Ann. Math. Statist., 35, 1704-1712.

Vandermonde determinants and expansions. Formulae for moments of elementary symmetric functions in s non-null characteristic roots. Moments of related roots of one matrix in terms of those of a second matrix.

880. Pillai, K.C.S. (1964): On the distribution of the largest of seven roots of a matrix in multivariate analysis. Biometrika, 51, 270-275.

Approximation of cumulative distribution function for the largest of seven characteristic roots. Pillai's approximate beta distribution. Tables of upper 1% and 5% points using approximation.

Pillai, K.C.S. (1964): See Gupta, S.S., Pillai, K.C.S. and Steck, G.P., #421.

881. Pillai, K.C.S. (1965): On the distribution of the largest characteristic root of a matrix in multivariate analysis. Biometrika, 52, 405-414.

Null distribution. Approximations to the distribution function of the largest of s roots. Reduction formulae and expansions for Vandermonde's determinant. Tables of 5 and 1 upper percentage points for s = 8, 9, 10.

882. Pillai, K.C.S. (1965): On elementary symmetric functions of the roots of two matrices in multivariate analysis. Biometrika, 52, 499-506.

The relation between the rth moment of the (s-i)th and ith e.s.f. of a matrix. Expressions for the first four moments of the (s-1)th e.s.f. The moments of the e.s.f.'s in the θ's from those of λ's where $\theta_i = \lambda_i/(1+\lambda_i)$ The moments of the second and the (s-2)th e.s.f.'s. Upper percentage points for the second e.s.f. by using the Pearson curve method. Tables of 5 and 1% points.

Pillai, K.C.S. (1965): See Gupta, S.S. and Pillai, K.C.S., #422.

Pillai, K.C.S. (1965): See Khatri, C.G. and Pillai, K.C.S., #598.

883. Pillai, K.C.S. (1966): Noncentral multivariate beta distribution and the moments of traces of some matrices. Multivariate Anal.-I, [Proc. 1st Internat. Symp., Krishnaiah, ed.], 237-251.

MV beta as the product of independent beta's in the linear case. Traces and moments of traces of some matrices. Moments of $W^{(2)}$, $V^{(2)}$ and $U^{(2)}$ (p = 2). Approximation to their distributions. Test of hypothesis λ = 0, against λ > 0 based on these statistics. Power. Largest root criterion. Test of the single non-null population canonical correlation being zero. Comparison of power of the tests includina largest root. Tables.

Pillai, K.C.S. (1966): See Khatri, C.G. and Pillai, K.C.S., #602.

884. Pillai, K.C.S. (1967): On the distribution of the largest root of a matrix in multivariate analysis. Ann. Math. Statist., 38, 616-617.

The distribution, under the hypothesis, of the largest root in the general case. Expression for the density as hypergeometric function of matrix argument.

885. Pillai, K.C.S. (1967): Upper percentage points of the largest root of a matrix in multivariate analysis. Biometrika, 54, 189-194.

Tables of upper percentage (5 and 1) points of the largest root θ_s for s = 14, 16, 18 and 20. Interpolation methods and errors. Approximations.

886. Pillai, K.C.S. and Gupta, A.K. (1967): On the distribution of the second elementary symmetric function of the roots of a matrix. Ann. Inst. Statist. Math., 19, 167-179.

The first four raw moments of second e.s.f. of s roots of a random matrix in terms of W determinant. Evaluation of the W determinants. Moments of the second, third and fourth e.s.f.'s. Approximations to the distribution of the second e.s.f. of s roots by a gamma type distribution. Example.

887. Pillai, K.C.S. and Jayachandran, K. (1967): Power comparisons of tests of two multivariate hypotheses based on four criteria. Biometrika, 54, 195-210.

Tests of hypotheses: (i) canonical correlation (ii) MANOVA Test criteria: (i) Roy's largest root, (ii) $U^{(p)}$, a constant times Hotelling's T_0^2, (iii) Pillai's $V^{(p)}$, (iv) Wilks' criterion. Exact noncentral cumulative distribution function for criteria (ii)-(iv) when p = 2. Approximation to largest root criterion for p = 2, 3, 4 in linear case. Comparison of power functions for four criteria with p = 2. Tables.

Pillai, K.C.S. (1967): See Khatri, C.G. and Pillai, K.C.S., #606.

888. Pillai, K.C.S. (1968): On the moment generating function of Pillai's $V^{(s)}$ criterion. Ann. Math. Statist., 39, 877-880.

Moment generating function of $V^{(s)}$ criterion in non-central case: (i) MANOVA situation, (ii) canonical correlation. Hypergeometric function of matrix argument. Zonal polynomials.

889. Pillai, K.C.S. and Gupta, A.K. (1968): On the noncentral distribution of the second elementary symmetric function of the roots of a matrix. Ann. Math. Statist., 39, 833-839.

Non-null moments of second elementary symmetric function of the roots of $|XX'-w\Sigma| = 0$. Zonal and Laguerre polynomials involving matrix argumen Approximation to noncentral distribution using moments. Tables showing adequacy of approximation.

890. Pillai, K.C.S. and Jayachandran, K. (1968): Power comparisons of tests of equality of two covariance matrices based on four criteria. Biometrika, 55, 335-342.

Test of $\Sigma_1 = \Sigma_2$ against one-sided alternatives. Four criteria studied: (i) Roy's largest root, (ii) Lawley-Hotelling, (iii) Pillai's, (iv) Wilks'. Exact non-null c.d.f. for (ii)-(iv) with $p = 2$. (Results for (i) in #897). Power comparisons for four criteria. Continuation of #887: Power comparison of Roy's criterion with (ii)-(iv) for hypotheses of independence (canonical correlation) and tests of equality of mean vectors (MANOVA).

Pillai, K.C.S. (1968): See Khatri, C.G. and Pillai, K.C.S., #609.

Pillai, K.C.S. (1968): See Khatri, C.G. and Pillai, K.C.S., #610.

891. Pillai, K.C.S. and Dotson, C.O. (1969): Power comparisons of tests of two multivariate hypotheses based on individual characteristic roots. Ann. Inst. Statist. Math., 21, 49-66.

Power comparisons for two cases (i) MANOVA, (ii) canonical correlation tests based on individual roots. Some recursive relations for the Vandermonde determinants. The distribution function of individual roots in the central case. $P\{x_i \leq x;\ m,\ n\} = 1-P\{x_{p-i+1} \leq 1 - x`\ n,\ m\}$. Non-null distribution of individual roots. Largest, smallest and median roots for the cases $p = 2, 3$. Tables of percentage points and power functions.

892. Pillai, K.C.S. and Gupta, A.K. (1969): On the exact distribution of Wilks's criterion. Biometrika, 56, 109-118. [Correction: 57 (1970), 225].

Modification of Schatzoff's exact distribution for Wilks' LR statistic. Repeated convolution with logX. Finite forms. Exact distribution of the criterion for $p = 3(1)6$. Tables of corrections for χ^2 for conversion into percentage points of the LR criterion. Three cases: (i) $\Sigma_1 = \Sigma_2$, (ii) MANOVA, (iii) independence of a p-set of variates and a q-set, for which the tables are useful, are pointed out.

893. Pillai, K.C.S. and Jouris, G.M. (1969): On the moments of elementary symmetric functions of the roots of two matrices. Ann. Inst. Statist. Math., 21, 309-320.

First two moments of $V_i^{(p)}$ and the three moments of $U_i^{(p)}$ when considered as the e.s.f.'s of the roots of random matrices. Moments of the second e.s.f. of the random matrix occurring in James' noncentral means distribution. Expressing the moments of $tr_i U$ for any λ in terms of the case when $\lambda = 0$. Tables of $a_{\kappa,\nu}$ for $k = 5,6,7,8$.

Pillai, K.C.S. (1969): See Khatri, C.G. and Pillai, K.C.S., #613.

894. Pillai, K.C.S. and Sugiyama, T. (1969): Noncentral distributions of the largest latent roots of three matrices in multivariate analysis. Ann. Inst. Statist. Math., 21, 321-327.

The density and distribution functions of four cases considered: (i) largest root of $|U_1 - (U_1+U_2)\lambda| = 0$, (ii) largest root of $|U_1 - fU_2| = 0$, (iii) canonical correlations, (iv) largest root of $|S_1 - gS_2| = 0$ in testing $\Sigma_1 = \Sigma_2$. Non-null distributions obtained in zonal polynomials.

895. Pillai, K.C.S., Al-Ani, S. and Jouris, G.M. (1969): On the distributions of the ratios of the roots of a covariance matrix and Wilks' criterion for tests of three hypotheses. Ann. Math. Statist., 40, 2033-2040.

Distribution of the ratios of the roots of a covariance matrix: $(e_1,\ldots,e_{p-1})$ where $e_i = w_i/w_p$, $p = 2, 3, 4$. Wilks' criterion and its non-null distribution in terms of Meijer's G-function and zonal polynomials. Three hypotheses: (i) $\Sigma_1 = \delta\Sigma_2$. Criterion is $w^{(p)} = |I-E|$ where E is a diagonal with $e_i = \delta f_i/(1+\delta f_i)$. Special case of $p = 2$. (ii) MANOVA with $w^{(p)} = |I-G|$, G being diagonal with g_i = ith root of $|S_1-g(S_1+S_2)| = 0$. The expression for the c.d.f. in $p = 2$ case. General density function given in a series involving Meijer's G and zonal polynomials. (iii) Canonical correlation case. Composite expression for all the three cases with a table giving the values of the parameters.

896. Pillai, K.C.S. (1970): On the noncentral distributions of the largest roots of two matrices in multivariate analysis. Essays Prob. Statist., (Roy Volume, Bose et al, eds.), 557-586.

Largest root criterion and power studies of the test involving the largest root in (i) canonical correlations case: (a) r_2^2 when ρ_1^2, ρ_2^2 are non-zero, (b) r_3^2 when one root is non-zero, (ii) MANOVA case: (a) e_2 when two roots ω_1, ω_2 of noncentrality matrix non-zero, (b) e_3, only one root non-zero. Tables of power. Upper percentage points of the largest root for $s = 11, 12$.

897. Pillai, K.C.S. and Al-Ani, S. (1970): Power comparisons of tests of equality of two covariance matrices based on individual characteristic roots. J. Amer. Statist. Assoc., 65, 438-446.

Exact noncentral distributions of individual characteristic roots for testing the hypothesis $\Sigma_1 = \Sigma_2$. *Expressions for the c.d.f. for the cases of* $p = 2, 3$. *Tables of power for the tests based on* g_1, g_2 $(p = 2)$, *and on* g_1, g_2, g_3 $(p = 3)$. *Various combinations of the roots of* $\Sigma_1\Sigma_2^{-1}$ *considered.*

Pillai, K.C.S. (1970): See Chattopadhyay, A.K. and Pillai, K.C.S., #154.

898. Pillai, K.C.S. and Jayachandran, K. (1970): On the exact distribution of Pillai's $V^{(s)}$ criterion. J. Amer. Statist. Assoc., 65, 447-454.

Moment generating function of $V^{(s)}$ *criterion in central case: weighted sum of determinants whose elements are incomplete gamma functions. Reduction of determinants in terms of single integrals or their products. Obtain c.d.f. by an inversion of m.g.f. Exact upper 5% and 1% points for* $V^{(3)}$ $(m = 1, 2, 3)$ *and* $V^{(4)}$ $(m = 0, 1)$.

899. Pillai, K.C.S. and Li, H.C. (1970): Monotonicity of the power functions of some tests of hypotheses concerning multivariate complex normal distributions. Ann. Inst. Statist. Math., 22, 307-318.

Complex analogue of the work of Anderson, Das Gupta and Mudholkar. Invariant tests based on symmetric group functions. Monotonicity of tests in complex MVN: linear hypothesis, equality of covariance matrices, independence of sets of variates.

Pillai, K.C.S. (1970): See Li, H.C., Pillai, K.C.S. and Chang, T.C., #725.

Pillai, K.C.S. (1971): See Chattopadhyay, A.K. and Pillai, K.C.S., #155.

Pillai, K.C.S. (1971): See Chattopadhyay, A.K. and Pillai, K.C.S., #156.

900. Pillai, K.C.S. and Jouris, G.M. (1971): Some distribution problems in the multivariate complex Gaussian case. Ann. Math. Statist., 42, 517-525.

Complex MVN. Complex multivariate beta distribution and independent beta variables. Noncentral distribution of $W^{(p)}$ *for testing equality of covariance matrices, MANOVA situation, independence of sets of variates. Distribution of* $V^{(2)}$ *and* $U^{(2)}$.

901. Pillai, K.C.S. and Nagarsenker, B.N. (1971): On the distribution of the sphericity test criterion in classical and complex normal populations having unknown covariance matrices. Ann. Math. Statist., 42, 764-767.

Distribution of $W = |S|/\{tr\ S/p\}^p$ for testing $\Sigma = \sigma^2 I$. Non-null distribution. Samples from real and complex MVN. Mellin transforms. Meijer's G-functions.

902. Pillai, K.C.S. and Young, D.L. (1971): An approximation to the distribution of the largest root of complex Wishart matrix. Ann. Inst. Statist. Math., 23, 89-96.

Complex Wishart in the canonical and central case. The c.d.f. of the largest root of a central complex Wishart matrix for $q = 2, 3, 4, 5$. Approximation in terms of linear combination of gamma functions. Tables of upper percentage points for $q = 2, 3, \ldots, 11$ for selected values of m. Application in time series analysis.

903. Pillai, K.C.S. and Young, D.L. (1971): On the exact distribution of Hotelling's generalized T_0^2. J. Multivariate Anal., 1, 90-107.

Laplace transform of $T_0^2 = n_2 U^{(s)}$. Vandermonde determinants and their expansions. Reduction formulae. Inversion of Laplace transforms. Exact non-null distribution of $U^{(p)}$. Tables of percentage points of $U^{(2)}$, $U^{(3)}$, $U^{(4)}$. Comparison with available approximations.

904. Pillai, K.C.S. (1972): The distribution of the characteristic roots of $S_1 S_2^{-1}$ under violations. Mimeo Ser. #278, Dept. Statist., Purdue U.

Joint distribution of the roots of the matrix $S_1 S_2^{-1}$ when S_1 is noncentral Wishart and S_2 is central Wishart. Expression for the density in a series involving (i) zonal polynomials, (ii) Laguerre polynomials. Special cases considered include one in which the noncentrality parameter is central Wishart (i.e., random).

Pillai, K.C.S. (1972): See Nagarsenker, B.N. and Pillai, K.C.S., #829.

905. Pillai, K.C.S. and Nagarsenker, B.N. (1972): On the distributions of a class of statistics in multivariate analysis. J. Multivariate Anal., 2, 96-114.

Noncentral distribution of the general function $Y = \phi_1^a \phi_2^b$

of the roots of a matrix where ϕ_1 *is the product of the roots* θ_i, ϕ_2 *is the product of* $(1-\theta_i)$. *Special cases considered are (i)* H_0: $\Sigma_1 = \Sigma_2$, *(iii) MANOVA, (iii) canonical correlations. The criteria that can be case in this form are Wilks'* Λ, *Wilks-Lawley U, modified LR statistic among others. Asymptotic distributions of the statistics are presented by use of Box's expansion of the characteristic functions. Characteristic functions given in each case.*

906. Pillai, K.C.S. and Sudjana (1972): On some distribution problems concerning the characteristic roots of $S_1S_2^{-1}$ under violations. Mimeo Ser. #287, Dept. Statist., Purdue U.

Generalization of the results on the distributions of various statistics: Hotelling's T_0^2, *Pillai's* $V^{(s)}$, *Wilks'* Λ *and the largest root. The moments of* T_0^2, *m.g.f. of* T_0^2 *are both obtained in a very general case. Zonal and Laguerre polynomials are used. The distribution theory developed in #904 for* $S_1S_2^{-1}$ *is used. The basic assumption made is that* S_1 *is noncentral Wishart with variance matrix* Σ_1, S_2 *is central Wishart with matrix* Σ_2.

Please, N.W. (1963): See Bartlett, M.S. and Please, N.W., #78.

907. Porebski, O.R. (1966): On the interrelated nature of the multivariate statistics used in discriminatory analysis. British J. Math. Statist. Psychol., 19, 197-214.

Review paper connecting the results due to Neyman-Pearson (LR), Fisher, Hotelling (canonical analysis), Wald (decision theoretic). Tests of significance are also examined and compared. These include Fisher's dispersion statistic, multiple correlation, Mahalanobis' distance, Hotelling's generalized t^2 *ratio, Wilks' LR, Hsu generalized correlation ratio. Table giving comparisons of the test procedures and statistics for various cases. Bibliographic.*

908. Posten, H.O. and Bargmann, R.E. (1964): Power of the likelihood-ratio test of the general linear hypothesis in multivariate analysis. Biometrika, 51, 467-480.

The model is $E[Y(Nxp)] = A(Nxk)\phi(kxp)$ *and the hypothesis* H_0: $C(n_hxk)\phi(kxp) = 0$. *Assumption is made of A being nonsingular. Relationship with Wilks-Lawley hypothesis. The moment function and the characteristic function of the LR statistic. Special cases of the linear and planar alternatives. Expansions of the LR criterion by characteristic function technique. Asymptotic distribution function in various powers of the sample size. Comparison with known results.*

Potthoff, R.F. (1958): See Roy, S.N. and Potthoff, R.F., #968.

909. Potthoff, R.F. and Roy, S.N. (1964): A generalized multivariate analysis of variance model useful especially for growth curve problems. Biometrika, 51, 313-326.

Generalization of the MANOVA model. Examples and applications of the model to real-life situations. Relation of the modified (growth-curve) model to the usual MANOVA model. Choice of the matrix G occurring in the transformation to usual MANOVA model. Numerical example. Simultaneous confidence bounds.

910. Poznyakov, V.V. (1971): On one representation of the multidimensional normal distribution function. Ukranian Math. J., 23, 562-566.

MVN distribution function and its partial derivative with respect to the correlation coefficient, ρ_{jk}. Differential difference equation relating the derivative of the n-dimensional c.d.f. with that of the two-dimensional c.d.f. and (n-2)-dimensional c.d.f.

Pratt, J.W. (1958): See Olkin, I. and Pratt, J.W., #850.

911. Press, S.J. (1967): On the sample covariance from a bivariate normal distribution. Ann. Inst. Statist. Math., 19, 355-361.

Distributional properties of sample covariance from a BVN. Series representations of density. Table of c.d.f. for even n and $\rho = 0$. Asymptotic distribution. Inference regarding ρ.

912. Press, S.J. (1967): Structured multivariate Behrens-Fisher problems. Sankhya, A, 29, 41-48.

Multivariate Behrens-Fisher problem for testing $\mu_1 = \mu_2$ vs. $\mu_1 \neq \mu_2$ with (i) intraclass correlation model for covariance, (ii) mean vectors of equal length, (iii) functionally related covariance matrices. LR tests. Scheffe-Bennett solution. Optimality property.

913. Press. S.J. (1969): The t-ratio distribution. J. Amer. Statist. Assoc., 64, 242-252.

Distribution of the ratio X/Y where (X,Y) are BV t-variates. The density and the distribution function. Graphs of the density for various combinations of the location and degrees of freedom. Significance points for 1,5,10,90,95 and 99 percent probabilities. Modes and asymptotic behaviour. Lack of moments.

Press, S.J. (1969): See Olkin, I. and Press, S.J., #855.

914. Press, S.J. (1972): Estimation in univariate and multivariate stable distributions. J. Amer. Statist. Assoc., 67, 842-846.

Estimation of the parameters of MVN by use of sample characteristic function.

915. Price, R. (1958): A useful theorem for nonlinear devices having Gaussian inputs. IRE Trans. Information Theory, 4, 69-72.

A necessary and sufficient condition for a vector random variable to be normal. Systems of partial differentail equations satisfied by the expectations of products of functions.

916. Prince, B.M. and Tate, R.F. (1966): Accuracy of maximum-likelihood estimates of correlation for a biserial model. Psychometrika, 31, 85-92.

Continuation of a paper by Hannan and Tate (#443). Moments of the ML estimators and tables of variances.

Putter, J. (1969): See Klotz, J. and Putter, J., #626.

917. Radcliffe, J. (1964): The construction of a matrix used in deriving tests of significance in multivariate analysis. Biometrika, 51, 503-504.

A matrix whose roots are equal to those of the residual matrix, to a certain order, is constructed. Lawley's approximate χ^2 *tests for the residual roots in canonical analysis, principal component analysis and factor analysis.*

918. Radcliffe, J. (1966): Factorizations of the residual likelihood criterion in discriminant analysis. Proc. Cambridge Philos. Soc., 62, 743-752.

Williams' results on several hypothetical discriminant functions and further study of the factorizations of the likelihood ratio. Approximate and exact factorizations. Removal of the effect of a set of assigned discriminant functions. Canonical forms. Split up of the total degrees of freedom into components. Application of Lawley's technique to approximate the distributions by that of χ^2*. Extension of Kshirasagar's results on the hypothetical discriminant functions in the space of the independent variables. More than one assigned discriminant function.*

919. Radcliffe, J. (1967): A note on an approximate factorization in discriminant analysis. Biometrika, 54, 665-668.

Demonstration of the result that the factors in the approximate factorization of residual likelihood correspond to LR criteria. Application of Lawley's result in this case to enforce agreement with χ^2 to a certain accuracy. Generalization of #918 so that all moments agree with those of χ^2 to this accuracy.

920. Radcliffe, J. (1968): A note on the construction of confidence intervals for the coefficients of a second canonical variable. Proc. Cambridge Philos. Soc., 64, 471-475.

Confidence intervals for the ratios of the coefficients of the second canonical variable (discriminant function) when the largest canonical correlation is near unity. Factorization of LR and its approximate distribution.

921. Radcliffe, J. (1968): The distributions of certain factors occurring in discriminant analysis. Proc. Cambridge Philos. Soc., 64, 731-740.

Extension of Kshirasagar's result to establish the distribution and independence of the factors (of the LR) in the case of s hypothetical discriminant functions. Analytical proof. MV beta and its decomposition. Expression of the factors in the case of several discriminant functions in terms of a MV beta and its components. Distributional properties when the independent variables are fixed and random. The former case corresponds to s hypothetical discriminant functions; the latter to s canonical variables. Distributions of the factors in the general case by a direct extension of Kshirasagar's results.

922. Radhakrishna, S. (1964): Discrimination analysis in medicine. Statistician, 14, 147-167.

A review paper on the applications and applicability of discriminant analysis in Medicine. A survey is made of the medical papers which have used this technique.

Radhakrishnan, R. (1972): See Sclove, S.L., Morris, C. and Radhakrishan, R., #1013.

923. Ramberg, J.S. (1972): Selection sample size approximations. Ann. Math. Statist., 43, 1977-1980.

Selection of population with largest mean when populations have a common known variance. Bechhofer procedure. Sample size approximations based on (i) Bonferroni in equality, (ii) Slepian inequality. Table comparing exact sample size with approximate ones. Extension to a multivariate problem: selecting the factor with the largest mean from a single MVN population with known Σ.

924. Rao, B.R. (1969): Partial canonical correlations. Trabajos Estadist., 20, 211-219.

Generalizes the concept of partial correlation between x_1 *and* x_2 *with respect to* x_3 *to partial canonical correlation between two sets of variables Y and Z with respect to a third set X. Multivariate linear model. MLE of partial canonical correlation.*

925. Rao, B.R., Garg, M.L. and Li, C.C. (1968): Correlation between the sample variances in a singly truncated bivariate normal distribution. Biometrika, 55, 433-436.

Truncation of one variable in BVN. Bivariate moments. Correlation coefficient, ρ^**, between sample variances,* m_{20} *and* m_{02}*, is related to correlation coefficient between variables in the truncated population and with coefficients of kurtosis of the marginal distributions. Tables of* ρ^* *for* $\rho = 0.50$ *and* 0.25 *for various points of truncation. Bounds for* ρ^*.

926. Rao, C.R. (1958): Some statistical methods for comparison of growth curves. Biometrics, 14, 1-17.

Analysis of repeated measurements data which are time dependent. Time metameter. Methods concerned with linearizing data and testing equality of regression coefficients for several groups: (i) Linearize response with respect to time and use ANOVA, (ii) comparison of average growth using Wilks' Λ*-criterion, (iii) large sample tests based on roots of determinantal equation. Analysis of individual growth curves: factor analysis or principal component analysis of dispersion matrix. Example.*

Rao, C.R. (1958): See Majumdar, D.N. and Rao, C.R., #739.

927. Rao, C.R. (1959): Some problems involving linear hypotheses in multivariate analysis. Biometrika, 46, 49-58.

Linear model: E(Y) = Aτ. Least squares with correlated variables. Dispersion matrix is unknown but estimate is available. Inference concerning τ: estimation, testing, confidence limits. Simultaneous confidence limits for a linear function of τ's. Application to growth curve analysis.

928. Rao, C.R. (1960): Multivariate analysis: an indispensible statistical aid in applied research. Sankhya, 22, 317-338.

Review paper emphasizing the applied aspects of various multivariate techniques. Bibliography.

929. Rao, C.R. (1961): Some observations on multivariate statistical methods in anthropological research. Bull. Inst. Internat. Statist., 38, 99-109.

Univariate vs. multivariate techniques. Discriminant function analysis and its application to anthropology. Analysis of growth curve data: time metameter, multivariate tests of significance. Canonical variate analysis vs. construction of indices.

930. Rao, C.R. (1962): A note on a generalized inverse of a matrix with applications to problems in mathematical statistics. J. Roy. Statist. Soc., B, 24, 152-158.

Properties of g-inverses. Computation of a g-inverse. Singular multivariate normal distribution. Property of χ^2 for quadratic forms. Unified approach to least squares theory.

931. Rao, C.R. (1962): Problems of selection with restrictions. J. Roy. Statist. Soc., B, 24, 401-405.

Determination of a linear function of $x_1, x_2, \ldots, x_p$ such that its correlation with y_1 is positive and a maximum subject to the condition that its correlations with $y_2, \ldots, y_q$ are all non-negative. Selection functions. Non-linear programming.

932. Rao, C.R. (1962): Use of discriminant and allied functions in multivariate analysis. Sankhya, A, 24, 149-154.

Classification into one of two groups. Prior test (used before discriminant function) to decide whether or not an observed specimen can be classified into one of the subsets. Possibility of belonging to unspecified group. Classification with parameters specified and estimated. Construction of linear functions of measurements which describe "size" and "shape".

933. Rao, C.R. and Varadarajan, V.S. (1963): Discrimination of Gaussian processes. Sankhya, A, 25, 303-330.

Equivalence and orthogonality properties of Gaussian processes. Gaussian measures on real Hilbert spaces. Likelihood ratios of two equivalent Gaussian measures on a real Hilbert space. Stability of Mahalanobis' D^2 as number of characters increases.

934. Rao, C.R. (1964): Sir Ronald Aylmer Fisher - the architech to multivariate analysis. Biometrics, 20, 286-300.

A review of literature on Fisher's contributions to multivariate analysis. Bibliographic.

935. Rao, C.R. (1964): The use and interpretation of principal component analysis in applied research. Sankhya, A, 26, 329-358.

A review of literature on principal components along with several interpretations. Some optimality properties. Weighted principal component analysis and analysis with restrictions on the individual variances. Instrumental variables. Applications to the interpretation of economic data. Size and shape factors. A justification for Jolicoeur-Mosimann interpretation. Anthropological applications. Relationship to factor analysis and multidimensional scaling. Application to physics. Clustering. Configurations of sets of points with respect to principal components. Tests of hypothesis on principal components. Use of principal component analysis in testing means.

936. Rao, C.R. (1965): The theory of least squares when the parameters are stochastic and its application to the analysis of growth curves. Biometrika, 52, 447-458.

Regression model $Y = AT+\varepsilon$ with $E(T) = \tau$, $Var(T) = e\Lambda$, $E(\varepsilon) = 0$, $Var(\varepsilon) = e\Sigma$, $Cov(T,\varepsilon) = 0$. The assumption is that there exists S an unbiased estimate of $Var\ Y = (A\Lambda A'+\Sigma)e$, which has a Wishart distribution independently of Y. The random components T, ε are both normal. Estimation of T and τ by LS method. Distributional properties. Tests on the appropriateness of the model and on the variance structure of Y. LR criteria and approximation by χ^2. Confidence intervals on T and τ. Application to growth curve problem. Covariance adjustment and Potthoff-Roy model of growth curves.

937. Rao, C.R. (1966): Discriminant function between composite hypotheses and related problems. Biometrika, 53, 339-345.

Mathematical formulation of Burnaby's paper on discriminant analysis and growth. Basic problem is the testing of the hypothesis that the distribution of X is dependent on a parameter θ which belongs to either a set θ_1 or θ_2 of parameter values. A number of procedures suggested: similar divisions, sufficient statistic, ancillary statistic and the method of maximum likelihood. Special cases dealt with are: (i) $E(X) = \alpha_i + B'\theta_i$, $V(X) = \Lambda$ under the hypothesis H_i: a_1, a_2, B are all known, (ii) same as (i) but with $V(X) = \Lambda_i$ under H_i, (iii) several composite hypotheses. The properties of the linear and quadratic functions studied under MVN assumption.

938. Rao, C.R. (1966): Covariance adjustment and related problems in multivariate analysis. Multivariate Anal.-I, [Proc. 1st Internat. Symp., Krishnaiah, ed.], 87-103.

Covariance adjustment in problems of (i) discriminant analysis, (ii) analysis of dispersion, (iii) Potthoff-Roy model of growth curves. Tests of hypotheses in various cases.

939. Rao, C.R. (1966): Characterization of the distribution of random variables in linear structural relations. Sankhya, A, 28, 251-260.

A characterization of MV normality: X = AF, X = BG, F, G being random, A, B not having any common colums nor multiples in each other, then X has to be MVN. Applications to econometric models and factor analytic models. Latent random variables.

940. Rao, C.R. (1967): Least squares theory using an estimated dispersion matrix and its application to measurement of signals. Proc. Fifth Berkeley Symp., 1, 355-372.

Least squares estimation and inference on τ in model $Y = X\tau + \varepsilon$ with unknown dispersion matrix of various structures. Structures of Σ_Y: (i) mixed model, (ii) factor analytic, (iii) autoregressive scheme, (iv) unknown arbitrary positive definite matrix. Test for adequacy of assumed Σ structure. Distributional properties. Covariance adjustment. Interval estimation of τ and simultaneous confidence bounds for function $P'\tau$. Characterization of the dispersion matrix for optimal character of least squares estimation. Estimation of phase and amplitude.

941. Rao, C.R. (1968): A note on a previous lemma in the theory of least squares and some further results. Sankhya, A, 30, 245-252.

Generalization of the paper #940: A NSC that for every parameter function or for a given subset, the BLUE with respect to Σ is the same as the BLUE with respect to $\Sigma = I$ or $\Sigma = \Sigma_0$ (known). Determination of the structure of Σ matrix.

942. Rao, C.R. (1969): A decomposition theorem for vector variables with a linear structure. Ann. Math. Statist., 40, 1845-1849.

Decomposition of a vector variable with linear structure into a sum of two independent components: (i) non-normal and with unique linear structure, (ii) multivariate normal with non-unique linear structure.

943. Rao, C.R. (1969): Recent advances in discriminatory analysis. J. Indian Soc. Agri. Statist., 21, 3-15.

Review of classification procedures based on discriminant functions. Sequential decision rules. Detecting specimen which may not belong to one of k specified subgroups. Constructing discriminant functions between composite hypotheses using (i) ancillary statistics, (ii) similar regions, (iii) method of maximum likelihood ratio.

944. Rao, C.R. (1969): Some characterizations of the multivariate normal distribution. Multivariate Anal.-II, [Proc. 2nd Internat. Symp., Krishnaiah, ed.], 321-328.

Characterization of multivariate normality on the basis of (i) regression, (ii) structural representation. Multivariate extension of Kagan-Linnik-Rao theorem for $n = 2$ and $n \geq 3$. Linear structural representation of random variables and characteristic property of MVN. Decomposition theorem (see #942).

945. Rao, C.R. (1970): Inference on discriminant function coefficients. Essays Prob. Statist., (Roy Volume, Bose et al, eds.), 587-602.

Sufficiency of a subset of variables as discriminators. Testing for equality of mean vectors after a covariance adjustment. Tests for discriminant function coefficients: (i) hypothesized set of coefficients (Fisher), (ii) given ratio of the coefficients of two variables, (iii) assigned ratios for coefficients of several variables. Discriminant functions in genetic selection.

946. Rao, C.R. (1971): Characterization of probability laws by linear functions. Sankhya, A, 33, 265-270.

Characterization of univariate and MV distributions by looking at the relation Z = AX. Identification of the distribution of X given that of Z.

947. Rao, C.R. (1972): Recent trends of research work in multivariate analysis. Biometrics, 28, 3-22.

A review of literature concerning discriminant analysi , test criteria, growth data. Archaeological seriation and characterization problems. Bibliographic.

948. Rao, J.N.K. (1958): A characterization of the normal distribution. Ann. Math. Statist., 29, 914-919. [Correction: 30 (1959), 610].

If $X_1,\ldots,X_n$ are such that for all functions g satisfying $g_i(x_1+a,\ldots,x_n+a) = g_i(x_1,\ldots,x_n)$, the sample mean $\bar{X}$ is independent of $g_1,\ldots,g_n$, then X is MVN.

Rao, M.M. (1961): See Krishnaiah, P.R. and Rao, M.M., #640.

949. Rao, M.M. (1964): Discriminant analysis. Ann. Inst. Statist. Math., 15, 11-24.

Discriminant analysis and classificatory procedures in the presence of trend. Elimination of the trend. LR criterion and its modification. Numerical example.

950. Rao, V.V.N. (1966): Distribution of some multivariate test statistics. J. Indian Statist. Assoc., 4, 195-201.

The distribution of the product of k beta variates by direct integration. Expression as a sum of infinite series. Application to distribution theory of (i) LR statistic (null) of MV linear hypothesis, (ii) the statistic for testing the hypothesis H_0: $\mu = 0$ in MV complex normal distribution.

Ratcliff, D. (1971): See Darroch, J.N. and Ratcliff, D., #214.

Rathie, P.N. (1970): See Mathai, A.M. and Rathie, P.N., #764.

Rathie, P.N. (1971): See Mathai, A.M. and Rathie, P.N., #767.

Rathie, P.N. (1971): See Mathai, A.M. and Rathie, P.N., #768.

Rathie, P.N. (1971): See Mathai, A.M. and Rathie, P.N., #769.

951. Rayner, A.A. and Livingstone, D. (1965): On the distribution of quadratic forms in singular normal variates. S. African J. Agri. Sci,, 8, 357-370.

MVN and conditional distributions using linear restrictions. Wishartness and χ^2 distribution with MVN variables. NSC for quadratic expressions to have Wishart distribution.

Rees, D.H. (1957): See Foster, F.G. and Rees, D.H., #305.

952. Regier, M.H. and Hamdan, M.A. (1971): Correlation in a bivariate normal distribution with truncation in both variables. Austral. J. Statist., 13, 77-82.

(X, Y) is BVN standardized with correlation ρ. Truncation of both X, Y to the left at a, b. Expression for ρ' the correlation in the truncated distribution in terms of ρ, a, b. Table for a = b.

Reinfurt, K.M. (1968): See Aitkin, M.A., Nelson, W.C. and Reinfurt, K.M., #18.

Reiter, S. (1961): See Ames, E. and Reiter, S., #24.

953. Reyment, R.A. (1962): Observations on homogeneity of covariance matrices in paleontologic biometry. Biometrics, 18, 1-11. [Correction: 18 (1962), 256].

Heterogeneous covariance matrices in testing H_0: $\mu_1 = \mu_2$ Modification of Mahalanobis' distance to accommodate $\Sigma_1 \neq \Sigma_2$, equal sample size case. Use of Bennett's solution. Example.

954. Reyment, R.A. (1969): A multivariate paleontological growth problem. Biometrics, 25, 1-8.

Heterogenity of variance matrices and testing for the equality of eigenvectors of two covariance matrices by extension of Anderson's results on testing for an eigenvector being equal to a specified vector. Example.

Riffenburgh, R.H. (1960): See Clunies-Ross, C.W. and Riffenburgh, R.H., #173.

Rizvi, M.H. (1966): See Alam, K. and Rizvi, M.H., #21.

Rizvi, M.H. (1966): See Krishnaiah, P.R. and Rizvi, M.H., #650.

955. Roberts, C. (1966): A correlation model useful in the study of twins. J. Amer. Statist. Assoc., 61, 1184-1190.

(X, Y) is BVN with mean (μ_1, μ_2), variances equal to σ^2 and covariance equal to $\rho\sigma^2$. Distribution of $Z = Min(X,Y)$ when $\mu_1 = \mu_2 = \mu$. Some properties of the distribution: moments of $\{(Z-\mu)/\sigma\}$. Distribution of $\{(Z-\mu)/\sigma\}^2$. Testing of hypothesis H_0: $\rho = \rho_0$ for large sample based on Z. Estimation of ρ and confidence interval. Unbiased estimation of ρ, case of $\mu_1 \neq \mu_2$. Example applying the results to data on twins.

956. Rosenbaum, S. (1961): Moments of a truncated bivariate normal distribution. J. Roy. Statist. Soc., B, 23, 405-408.

Truncation of BVN (singly) in both variables. Moments and estimation by iterative procedure. Example.

Rosenblatt, H.M. (1957): See Kullback, S. and Rosenblatt, H.M., #693.

957. Ross, J. and Weitzman, R.A. (1964): The twenty-seven percent rule. Ann. Math. Statist., 35, 214-221.

Estimation of the coefficient of correlation by the proportion (or number) in symmetrically placed corner regions ignoring the middle section. ML method. Asymptotic variance formula. Table of values of the variances. Relative efficiency. Cost considerations.

Rothenberg, T. (1969): See Groves, T. and Rothenberg, T., #404.

Roux, J.J.J. (1970): See Steyn, H.S. and Roux, J.J.J., #1091.

958. Roux, J.J.J. (1971): A characterization of a multivariate distribution. South African Statist. J., 5, 27-36.

Extension of work of Seshadri and Patil to (i) multivariate random variables, (ii) matrix valued random variables. Characterization of joint distribution in terms of conditional and marginal distributions. Moment generating functions. Multivariate Dirichlet and multivariate normal distributions.

Roux, J.J.J. (1972): See Steyn, H.S. and Roux, J.J.J., #1092.

959. Roy, J. (1958): Step-down procedure in multivariate analysis. Ann. Math. Statist., 29, 1177-1187.

Sequential testing. Stepwise procedures for testing multivariate situations in which variables can be ordered in importance a priori and arranged in descending order. Each component hypothesis associated with well-known univariate distribution. Component hypotheses are independent. Probability level easily determined. Procedure applied to (i) problems in MANOVA, (ii) testing $\Sigma = \Sigma_0$. Simultaneous confidence bounds based on step-down techniques.

Roy, J. (1959): See Roy, S.N. and Roy, J., #972.

Roy, J. (1959): See Roy, S.N. and Roy, J., #973.

960. Roy, J. (1960): Tests for independence and symmetry in multivariate normal populations. Sankhya, 22, 267-278.

Asymptotic series expansion of LR criterion in terms of χ^2 probability integrals. Tables of coefficients for various tests. Tests of independence and symmetry (Mauchly, Votaw, Wilks) in a single MVN population. Tests of equality of variances and covariances. Tests of compound symmetry. Tests of mutual independence. Tests of equality of means, variances and covariances.

961. Roy, J. (1960): Non-null distribution of the likelihood ratio in analysis of dispersion. Sankhya, 22, 289-294.

Non-null distribution of Wilks' LR criterion for MANOVA. Decomposition of criterion. Approximation. Evaluation of power of MANOVA test when alternative hypothesis is of rank one. Gamma series expansion. Adequate for large error d.f. and small noncentrality parameter. Table of power given for $n = 200$, $\alpha = .05$, $m = 2, 3$, $p = 1, 2, 3, 4$ and $\delta^2 = 0(1)8$.

962. Roy, J. and Murthy, V.K. (1960): Percentage points of Wilks' L_{mvc} and L_{vc} criteria. Psychometrika, 25, 243-250.

LR criterion for testing equality of (i) means, variances and covariances, (ii) variances and covariances in MVN when n is large. Asymptotic expansion in terms of weighted χ^2 variables. Upper 1% and 5% points for $p = 4, 5, 6, 7$ and $n = 25(5)60(10)100$.

963. Roy, J. (1966): Power of the likelihood-ratio test used in analysis of dispersion, Multivariate Anal.-I, [Proc. 1st Internat. Symp., Krishnaiah, ed.], 105-127.

Power function of Wilks' LR criterion for the general multivariate linear hypothesis when alternative hypothesis is of unit rank. Exact expression for p = 2. Approximations to power:: (i) expansion in gamma series, (ii) using Jacobi polynomials. Tables comparing approximations and exact values of power for p = 2 and α = .05.

964. Roy, S.G. and Mukherjee, G.D. (1964): Use of Mahalanobis' D^2-statistic in phase-discrimination in desert locust males. Sankhya, B, 26, 237-252.

Application of Mahalanobis' D^2 to classify desert locusts into one of two phases. Tables of dispersion matrices and their inverse.

965. Roy, S.N. and Gnanadesikan, R. (1957): Further contributions to multivariate confidence bounds. Biometrika, 44, 399-410. [Correction: 48 (1961), 474].

Simultaneous confidence bounds on the roots of (i) a single variance matrix Σ, (ii) two dispersion matrices Σ_1, Σ_2, by looking at $\Sigma_1\Sigma_2^{-1}$, (iii) regression matrix of p-variate set on a q-variate set, (iv) matrix connected with M V linear hypothesis on means. General set up in M V linear hypothesis cases. Example. Discriminant analysis. Truncation of the matrices and components of the vectors. An optimal property of the simultaneous confidence bounds.

966. Roy, S.N. and Bargmann, R.E. (1958): Tests of multiple independence and the associated confidence bounds. Ann. Math. Statist., 29, 491-503.

Stepdown multiple correlations and their independence. Simultaneous inferences concerning these correlations. Relation to the LR test. Simultaneous confidence bounds on the regression matrix. Stepdown sets of canonical correlations and independence. Test and its relation to LR. SCB associated with the sets of variates.

967. Roy, S.N. and Gnanadesikan, R. (1958): A note on "Further contributions to multivariate confidence bounds". Biometrika, 45, 581.

A note on the application of the SCB techniques to include the truncation of the C matrix. Thus an SCB has preassigned probability 1-α concerning $(2^u-1)(2^r-1)$

confidence statements obtained by all possible truncations of the M and C matrices.

968. Roy, S.N. and Potthoff, R.F. (1958): Confidence bounds on vector analogues of the "ratio of means" and the "ratio of variances" for two correlated normal variates and some associated tests. Ann. Math. Statist., 29, 829-841.

Ratio of the means and variances in BVN and MVN cases. Confidence bounds on (i) μ_1/μ_2, σ_1^2/σ_2^2 in the BV case, (ii) functions suitably defined in the MVN, to be interpreted as ratios. Truncation and its role in SCB. Union Intersection principle. Central t, T^2 and canonical correlation.

969. Roy, S.N. (1959): A note on confidence bounds connected with the hypothesis of equality of two dispersion matrices. Calcutta Math. Soc. Golden Jubilee Commem. Volume 2, 329-332.

Characteristic roots of $S_1S_2^{-1}$ and $\Sigma_1\Sigma_2^{-1}$. The result $\lambda_1 c_{max}(S_1S_2^{-1}) \geq$ all $c(\Sigma_1\Sigma_2^{-1}) \geq \lambda_2 c_{min}(S_1S_2^{-1})$ extended to matrices obtained by eliminating variates or sets of variates. Proof uses partitioned matrices.

970. Roy, S.N. and Gnanadesikan, R. (1959): Some contributions to ANOVA in one or more dimensions: I. Ann. Math. Statist., 30, 304-317.

Univariate ANOVA discussed with references to Models I and II. Definition of parametric functions. SCB and individual confidence bounds.

971. Roy, S.N. and Gnanadesikan, R. (1959): Some contributions to ANOVA in one or more dimensions: II. Ann. Math. Statist., 30, 318-340.

MANOVA. Models I and II introduced. Model I is tested using maximum root statistic. Testable, quasi-independent hypotheses. NSC for these conditions to hold. Confidence bounds. MV components of variance. Linear estimation and testing of linear hypotheses. NSC for estimability. Pseudo-Wishart and NS conditions for the quadratic expressions to be pseudo-Wishart and independent. Connection between the analyses under models I and II for the restricted k-way classification. Tests of hypotheses in the Model II. Maximum root criterion. Confidence bounds.

972. Roy, S.N. and Roy, J. (1959): A note on a class of problems in normal multivariate analysis of variance. Ann. Math. Statist., 30, 577-581.

MANOVA with E(X) = Aξ, A known and ξ a matrix of unknown constants. Hypothesis H: $\xi = B\eta$, B is given, η is unknown matrix. Complete testability and NSC for this. Construction of testable hypotheses implied by H. Criterion based on the roots of $S_2S_1^{-1}$ where S_1, S_1+S_2 are error matrices under the model and under H, respectively.

973. Roy, S.N. and Roy, J. (1959): A note on testability in normal model I, ANOVA and MANOVA. Calcutta Math. Soc. Golden Jubilee Commem. Volume 2, 333-339.

Testing hypothesis that matrix of means satisfies a given linear relation. Test based on largest characteristic root of $S_2S_1^{-1}$ where S_1 and S_1+S_2 are error matrices under model and hypothesis, respectively.

974. Roy, S.N. and Cobb, W. (1960): Mixed model variance analysis with normal error and possibly non-normal other random effects, II. The multivariate case. Ann. Math. Statist., 31, 958-968.

MV mixed model with random block effects. Confidence bounds on the maximum and minimum roots of the variance of the block effects for MVN case. The non-normal case. Marginal and conditional quantiles. Confidence bounds in the bivariate case on the maximum and minimum roots.

975. Roy, S.N., Greenberg, B.G. and Sarhan, A.E. (1960): Evaluation of determinants, characteristic equations and their roots for a class of patterned matrices. J. Roy. Statist. Soc., B, 22, 348-359.

Patterned matrices applicable in design and other situations.

976. Roy, S.N. and Mikhail, W.F. (1961): On the monotonic character of the power functions of two multivariate tests. Ann. Math. Statist., 32, 1145-1151.

Monotonicity of the power function of the largest root test for MANOVA and independence between sets of variates. The monotonicity is in each noncentrality parameter separately and thus generalizes previous results.

977. Roy, S.N. (1962): A survey of some recent results in normal multivariate confidence bounds. Bull. Inst. Internat. Statist., 39, 405-422.

Simultaneous confidence bounds. Construction of confidence bounds using Roy's Union-Intersection principle. Bounds related to (i) a single dispersion matrix and two dispersion matrices, (ii) multivariate linear hypothesis, (iii) independence of sets of variates.

978. Roy, S.N. and Gnanadesikan, R. (1962): Two-sample comparisons of dispersion matrices for alternatives of intermediate specificity. Ann. Math. Statist., 33, 432-437.

Union-Intersection principle. Tests of $\Sigma_1 = \Sigma_2$ vs. alternatives expressed in terms of characteristic roots of $\Sigma_1\Sigma_2^{-1}$. Inference procedures based on similar regions. Simultaneous confidence bounds.

Roy, S.N. (1964): See Potthoff, R.F. and Roy, S.N., #909.

979. Roy, S.N. and Srivastava, J.N. (1964): Hierarchical and p-block multiresponse designs and their analysis. Contrib. Statist., (Mahalanobis Volume, Rao, ed.), 419-428.

Definition of hierarchical multiresponse (HM) designs. HM designs with p-clock systems. Testability conditions of HM designs. Tests concerning treatment effects using step-down procedures.

980. Roy, S.N. and Mikhail, W.F. (1970): The admissibility of the largest characteristic root test for the normal multivariate linear hypotheses. Essays Prob. Statist., (Roy Volume, Bose et al, eds.), 621-630.

Largest characteristic root test of MANOVA is admissible among the class of similar tests invariant under the group of scale transformations. Union-Intersection principle. Quasi-similar and similar regions. Most powerful quasi-similar tests in the class of Neyman structure tests.

981. Ruben, H. (1960): Probability content of regions under spherical normal distributions I. Ann. Math. Statist., 31, 598-618.

A review article outlining the problem of integration. Probability contents of (i) halfspace, (ii) central and noncentral hyperspheres, (iii) hyperspherical cone and cylinder, (iv) variety of revolution of dimension (n-1) and species p, (v) ellipsoid, (vi) regular simplex. Applications to derivation of some standard noncentral distributions. Bibliographic.

982. Ruben, H. (1960): Probability content of regions under spherical normal distributions II: The distribution of the range in normal samples. Ann. Math. Statist., 31, 1113-1121. [Correction: 32 (1961), 620].

Expressing the distribution function of the range of n observations from N(0,1) as a multiple integral of (n-1) variables over a parallelotype. Determination of the angles at the vertices. Evaluation of the probability content of such regions by method of sections. Special cases of $n = 2, 3$. Expansion for the distribution function, as a series involving the even moments of truncated normal distribution. Bibliographic.

983. Ruben, H. (1961): Probability content of regions under spherical normal distributions III: The bivariate normal integral. Ann. Math. Statist., 32, 171-186.

Evaluation of $L \equiv P\{X > x_0, Y > y_0\}$, in a standard BVN distribution. Representation of the integral in terms of independent N(0,1) variables. Geometrical description of the various configurations (32 in all). Expansion for the integral L and of a related function W of a noncentral t with one degree of freedom. Mill's ratio. Continued fractions and expansions for L and W.

984. Ruben, H. (1961): An asymptotic expansion for a class of multivariate normal integrals. J. Austral. Math. Soc., 2, 253-264.

Integral $I \equiv P\{X > h\}$, when X is MVN with mean zero and $\Sigma = \{I\rho + J(1-\rho)\}$. Generalization of Mill's ratio. Expression of the integral in terms of independent N(0,1) integrals over a polyhedral half-cone. Reduction to lower dimensions in terms of k-functions. Series expansion: Accuracy of the asymptotic expansion.

985. Ruben, H. (1961): On the numerical evaluation of a class of multivariate normal integrals. Proc. Roy. Soc. (Edinburgh), A, 65, 272-281.

Probability of $\{X > h\}$, for the case of standard equicorrelated X, is expressed as the product of $f(h,\ldots,h)$ the density at h, and an infinite series in h. Geometrical interpretation of the coefficients of the series. Evaluation of the constants occurring in the coefficients. Expression in terms of measures of regular hyperspherical simplices. Special cases of $n = 1, 2$.

986. Ruben, H. (1962): Probability content of regions under spherical normal distributions IV: The distribution of homogeneous and non-homogeneous quadratic functions of normal variables. Ann. Math. Statist., 33, 542-570.

Distribution of homogeneous and non-homogeneous quadratic forms from a MVN distribution. Evaluation of the integrals. Expansions in terms of central and noncentral χ^2 distribution functions. Accuracy of the expansions.

987. Ruben, H. (1964): An asymptotic expansion for the multivariate normal distribution and Mill's ratio. J. Nat. Bur. Standards, Sect. B, 68, 3-11.

MV analogue of Mill's ratio, i.e., $F(a,M)/f(a,M)$. $F = P\{X > a\}$, f is the p.d.f. evaluated at $X = a$. Here X is MVN with mean zero and variance equal to M. Asymptotic expansion similar to that for the univariate case. Properties of the expansion and its possible generalizations. Special case of equicorrelated standard MVN.

988. Ruben, H. (1966): On the simultaneous stabilization of variances and covariances. Ann. Inst. Statist. Math., 18, 203-210.

Limit theorems for vector-valued random variables and their applications. Variance stabilizing formulae for the MV case. That is, if t has variance matrix $\Sigma(\theta)$, θ being some set of parameters, then there exists a transformation on t such that transformed variables have variance matrix which is independent of θ. The theorem is proved for the case when $(t_n-\theta)/\sqrt{n}$ converges in distribution to $N(0,\Sigma)$. Applications suggested for MANOVA.

989. Ruben, H. (1966): Some new results on the distribution of the sample correlation coefficient. J. Roy. Statist. Soc., B, 28, 513-525.

Transformation on r as $\tilde{r} = r/(1-r^2)^{\frac{1}{2}}$. A representation of $\tilde{r} = \{Z+\tilde{\rho}\chi_{n-1}\}/\chi_{n-2}$ where Z is $N(0,1)$ and χ is a chi variable with appropriate degrees of freedom and $\tilde{\rho} = \rho/(1-\rho^2)^{\frac{1}{2}}$. Hence a derivation of the distribution of r in normal samples both when $\rho = 0$ and $\rho \neq 0$. Alternative short proof for the latter case. $P\{\tilde{r} > 0\} = P\{t_{n-1} > -(n-1)^{\frac{1}{2}}\rho\}$. Moments of $\tilde{r}$. Approximate distributions for r. Examination of the approximation.

Rubin, H. (1962): See Olkin, I. and Rubin, H., #853.

Rubin, H. (1964): See Olkin, I. and Rubin, H., #854.

990. Rubin, H. (1969): Decision theoretic approach to some multivariate problems. Multivariate Anal.-II, [Proc. 2nd Internat. Symp., Krishnaiah, ed.], 507-513.

A general paper recommending the use of decision theoretic approach to MV analysis. Specifically, in the case of regression, structural, discriminant, factor and data analytic methods.

Rubin, J. (1967): See Friedman,H.P. and Rubin, J., #315.

Russell, J.S. (1968): See Horton, I.F., Russell, J.S. and Moore, A.W., #478.

991. Samiuddin, M. (1970): On a test for an assigned value of correlation in a bivariate normal distribution. Biometrika, 57, 461-464.

A new transformation of r *for* $\rho \neq 0$ *which yields a Student's* t *distribution. Non-null distribution of* r *and the moments of the new statistic. Comparisons with standard results. Application to confidence interval estimation of* ρ.

Samson, P. (1959): See Pillai, K.C.S. and Samson, P., #878.

Sanghvi, L.D. (1968): See Balakrishnan, V. and Sanghvi, L.D., #67.

992. Sankaran, M. (1958): On Nair's transformation of the correlation coefficient. Biometrika, 45, 567-571.

A transformation of r *as* $x = (r-\rho)/(1-\rho r)$. *Asymptotic normality of* $\sin^{-1} x$ *by moments. Comparison of the exact and approximate probability integrals of* r. *Confidence interval for* ρ.

Sarhan, A.E. (1959): See Greenberg, B.G. and Sarhan, A.E., #399.

Sarhan, A.E. (1960): See Greenberg, B.G. and Sarhan, A.E., #400.

Sarhan, A.E. (1960): See Roy, S.N., Greenberg, B.G. and Sarhan, A.E., #975.

993. Sarmanov, O.V. and Zakharov, V.K. (1960): Maximum coefficients of multiple correlation. Soviet Math. Dokl., 1, 51-53.

Concept of maximum coefficient of correlation between random vectors. Characterization of mutual independence of a set of random variables.

994. Sato, S. (1962): A multivariate analogue of pooling of data. Bull. Math. Statist., 10, 61-76.

Two sample problem. Pooling of estimates of the mean after a preliminary test. Distribution of the pooled estimate, its mean and variance. Two cases dealt with are Σ (equal) known or unknown.

Sato, S. (1962): See Asano, C. and Sato, S., #50.

995. Savage, I.R. (1962): Mill's ratio for multivariate normal distributions. J. Res. Nat. Bur. Standards, Sect. B, 66, 93-96.

$P\{X > a\}$ when X is MVN with $E(X) = 0$, $V(X) = M$. Inequality for this probability in terms of $f(a)$, the p.d.f. at $X = a$ and its application to special cases.

996. Saw, J.G. (1970): The multivariate linear hypothesis with nonnormal errors and a classical setting for the structure of inference in a special case. Biometrika, 57, 531-535.

Linear hypothesis in MV case when observations are proportional to errors. Distributions in a general non-normal case. Special case of normality and Wishartness. Fraser's progression model. Structure of inference in a special case and a classical setting for the same.

Saw, J.G. (1970): See Hughes, D.T. and Saw, J.G., #480.

Saw, J.G. (1970): See Hughes, D.T. and Saw, J.G., #481.

997. Saw, J.G. (1971): A conservative test for the rank of a noncentrality matrix. Bounds on the distribution of a product of latent roots. Tech. Rep. #24, Dept. Statist., U. Florida.

The noncentrality parameter of a Wishart and the test for its rank being less than or equal to (m-k) for a given k and dimensionality m. LR criterion. Upper bound for the probability of rejecting H_0. Application of the asymptotic distribution of the LR statistic. Lower bound for this probability. Conservativeness of the test.

998. Saw, J.G. (1972): The occurrence of a random orthogonal matrix in multivariate distribution theory. Tech. Rep. #39, Dept. Statist., U. Florida.

The demonstration of the orthogonal matrix as a naturally occurring random variable and the actual distribution is determined as a marginal of a joint distribution. Example of Wishart and latent root distributions.

Saw, J.G. (1972): See Hughes, D.T. and Saw, J.G., #482.

999. Saxena, A.K. (1966): On the complex analogue of T^2 for two populations. J. Indian Statist. Assoc., 4, 99-102.

LR test of equality of means in two samples. Complex normal case. Expressing the statistic in terms of T^2. Distribution of the T^2 as central (noncentral) F in the null (non-null) cases.

1000. Saxena, A.K. (1967): Invariance of the power function of a multivariate test. J. Indian Statist. Assoc., 5, 66-67.

Application of Lehmann's theorem to demonstrate that Giri's test of a multivariate hypothesis concerning $\Gamma = \Sigma^{-1}\mu$ is invariant in power. Uniformly most powerful invariant similar test.

1001. Saxena, A.K. (1967): A note on classification. Ann. Math. Statist., 38, 1592-1593.

This note extends the usual problem of classification to the complex case. Bayes solution. Complex analogue of Hotelling's T^2. Σ common.

1002. Saxena, A.K. (1969): On an application of Matusita's distance function. J. Indian Statist. Assoc., 7, 73-76.

Complex analogue of Matusita's distance function in the case of complex Gaussian distribution. Application to classificatory problem in the complex, equal variance matrix case.

Saxena, R.K. (1969): See Mathai, A.M. and Saxena, R.K., #762.

1003. Schatzoff, M. (1966): Sensitivity comparisons among tests of the general linear hypothesis. J. Amer. Statist. Assoc., 61, 415-435.

Definition of expected significance level (ESL) of a test: the expected value of the observed significance level under a simple alternative hypothesis. Using ESL as a criterion for comparing sensitivities of competing test statistics. Comparison of tests of general linear hypothesis: (i) Wilks' Λ, (ii) Hotelling's T_0^2, *(iii) Roy's largest root. Tables and graphs of ESL. Monte Carlo studies.*

1004. Schatzoff, M. (1966): Exact distributions of Wilks's likelihood ratio criterion. Biometrika, 53, 347-358. [Correction: 54 (1967), 688].

Density and distribution function of Wilks' Λ for MANOVA situation. Closed form representations when p, number of variates, or q, hypothesis degrees of freedom, are even. Tables of factors needed for converting χ^2 *percentiles to exact percentiles of a logarithmic function of Λ. Example of us of tables in MANOVA.*

1005. Scheur, E.M. and Stoller, D.S. (1962): On the generation of normal random vectors. Technometrics, 4, 278-281.

Generation of a random vector from a MVN (0,Σ) population. Methods used: (i) linear transformations, (ii) conditional distributions.

1006. Schull, W.J. and Kudo, A. (1962): Certain multivariate problems arising in human genetics. Bull. Math. Statist., 10, 77-88.

Multivariate regression analysis with q populations where each observation vector consists of k main variables and n concomitant ones. Regression between main variables and concomitant ones. Equality of dispersion matrices. Equality of matrices of regression coefficients. Techniques illustrated by human genetics data.

Schull, W.J. (1964): See Ito, K. and Schull, W.J., #491.

Schwartz, R. (1965): See Kiefer, J. and Schwartz, R., #622.

1007. Schwartz, R. (1966): Fully invariant proper Bayes tests. Multivariate Anal.-I, [Proc. 1st Internat. Symp., Krishnaiah, ed.], 275-284.

Construction of a priori distributions which yield invariant tests. Transformation groups. Stein's representation of the probability density of the maximal invariant as an integral over the appropriate transformation group. MANOVA tests. Proper and improper Bayes tests.

1008. Schwartz, R. (1967): Locally minimax tests. Ann. Math. Statist., 38, 340-359.

Property of local minimaxity. MANOVA problem and invariance. Power function of a fully invariant test. Acceptance regions based on the trace of a matrix in MANOVA. Local minimax properties of this test. Local minimax properties of invariant tests of independence of sets of variates (more than two). G-admissibility and the power of G-invariant tests.

1009. Schwartz, R. (1967): Admissible tests in multivariate analysis of variance. Ann. Math. Statist., 38, 698-710.

Admissibility of fully invariant tests. Stein's theorem on testing problems for exponential families. Application and corollaries to Stein's theorem. Exponential nature of MANOVA problem. Restricted alternatives. Characterization of all admissible fully invariant tests: application of exponential method.

1010. Schwartz, R. (1969): Invariant proper Bayes tests for exponential families. Ann. Math. Statist., 40, 270-283.

Characterization of invariant Bayes tests. Stein's representation. Applications to MANOVA tests of independence of sets of variates and equality of proportional covariance matrices.

1011. Sclove, S.L. (1970): Some remarks on normal multivariate regression. Ann. Inst. Statist. Math., 22, 319-326.

MV normal multiple regression. Extension of Stein's results. Definition of a loss function. The minimax property of the MLE for this loss function. Invariant estimator. Prediction functions and their minimax nature. Example.

1012. Sclove, S.L. (1971): Improved estimation of parameters in multivariate regression. Sankhya, A, 33, 61-66.

MV linear regression and MLE for B. Insensitivity to Σ. Equivalence to estimation of mean vectors. Canonical reduction. Estimator dominating the MLE in the sense of expected loss. Preliminary test estimator.

1013. Sclove, S.L., Morris, C. and Radhakrishnan, R. (1972): Non-optimality of preliminary test estimators for the mean of a multivariate normal distribution. Ann. Math. Statist., 43, 1481-1490.

Estimation of μ in the decision theoretic context. Quadratic loss. Preliminary test estimators when $\Sigma = \sigma^2 I$, $\sigma^2 B$, Σ general. Comparison with the usual estimator in terms of risk to demonstrate it does not dominate nor is it dominated. James-Stein estimator dominates PTE. Estimation under the subvector being assigned, μ belonging to a subspace, general linear hypothesis.

1014. Scott, A. (1967): A note on conservative confidence regions for the mean of a multivariate normal. Ann. Math. Statist., 38, 278-280. [Correction: 39 (1968), 2161].

The proof of Dunn's conjecture, namely $P\{|Z_i| \le c_i, i = 1,\ldots,m\} \ge \Pi P\{Z_i \le c_i\}$ in the general case of any m and any Σ. The cases considered are (i) diagonal elements of Σ known, (ii) diagonal elements estimated. In (i) $Z_i = (X_i - \mu_i)/\sigma_i$ and in (ii) $Z_i = (X_i - \mu_i)/s_i$.

1015. Scott, A. and Symons, M.J. (1971): On the Edwards and Cavalli-Sforza method of cluster analysis. Biometrics, 27, 217-219.

Grouping of N observations of ν variates. Method requiring $(2^{\nu}-2)\binom{N}{\nu}$ partitions instead of Edwards-Cavalle-Sforza result using $(2^{N-1}-1)$ partitions.

1016. Scott, A. and Symons, M.J. (1971): Clustering methods based on likelihood ratio criteria. Biometrics, 27, 287-297.

Cluster analysis by the estimation of a parameter γ of n components which assigns an observation to the group from which it comes. ML method. Classification as one of estimation of γ and of the clusters: (i) equal variances for all clusters, (ii) unequal covariances. Bayes estimation with diffuse priors. Numerical example. Comparison with common clustering procedures. Drawbacks of these procedures.

1017. Seal, H.L. (1967): Studies in the history of probability and statistics XV. The historical development of the Gauss linear model. Biometrika, 54, 1-24.

Historical development of MV analysis with the discussion of K. Pearson's and R. A. Fisher's contributions.

Sedransk, J. (1971): See Brogan, D.R. and Sedransk, J., #125.

1018. Seely, J. (1971): Quadratic subspaces and completeness. Ann. Math. Statist., 42, 710-721.

Quadratic subspaces, their properties. MVN with zero mean and variance matrix having the structure $\Sigma k_j V_j$, V known and k unknown. Estimable functions and quadratic subspaces. Complete sufficient statistics in the non-zero mean case. Example: Mixed linear model.

1019. Seely, J. (1972): Completeness for a family of multivariate normal distributions. Ann. Math. Statist., 43, 1644-1647.

NSC for a special family of MVN distributions to admit a complete sufficient statistic. The family is taken to have zero mean and variance matrix of the form $\Sigma k_i V_i$ where k_i are unknown and V_i are known matrices. This paper establishes a converse of the result proved in an earlier one. Representations for locally best unbiased estimators. Principle of invariance and its application to the case when expectation is not zero. MINQUE.

1020. Seshadri, V. (1966): A characteristic property of the multivariate normal distribution. Ann. Math. Statist., 37, 1829-1831.

A characterization of MVN: If X, Y are independent and if conditional of X given X+Y is MVN with expectation C(X+Y), variance V, then X, Y are both MVN. Conditions on C, V given.

Severo, N.C. (1960): See Zelen, M. and Severo, N.C., #1187.

1021. Shah, B.K. (1970): Distribution theory of a positive definite quadratic form with matrix argument. Ann. Math. Statist., 41, 692-697.

Laguerre polynomial representation of the distribution. Some integrals involving zonal and Laguerre polynomials. The moment generating function.

1022. Shakun, M.F. (1965): Multivariate acceptance sampling procedures for general specification ellipsoids. J. Amer. Statist. Assoc., 60, 905-913.

Known covariance case. Single sampling acceptance plans. Determination of n and the acceptance number c.

Sheth, R.J. (1966): See Parikh, N.T. and Sheth, R.J., #863.

1023. Shimizu, R. (1962): Characterization of the normal distribution II. Ann. Inst. Statist. Math., 14, 173-178.

Proves characterization theorem concerning univariate normal: Let $x_1, x_2, \ldots, x_n$ be a sample from population with distribution function F(x) and assume second moment exists. If $a_1x_1 + \ldots + a_nx_n$ has the same distribution as x_1, then F(x) is normal. Extended to multivariate normal.

1024. Shine, L.C., II (1972): A note on McDonald's generalization of principal components analysis. Psychometrika, 37, 99-101.

McDonald's (#780) extremum criterion applied to principal component analysis for groups of variables. Total variance of original variables being channeled through groups of variables.

Shrikhande, S.S. (1970): See Olkin, I. and Shrikhande, S.S., #856.

1025. Sidak, Z. (1967): Rectangular confidence regions for the means of multivariate normal distributions. J. Amer. Statist. Assoc., 62, 626-633.

Probability inequalities for MVN. Rectangular confidence regions for the mean vector when (i) variances are known, (ii) variances are unknown but equal. Comparison of rectangular and ellipsoidal regions.

1026. Sidak, Z. (1968): On multivariate normal probabilities of rectangles: their dependence on correlations. Ann. Math. Statist., 39, 1425-1434.

Probabilities of rectangles in MVN. Two sided analogue of Slepian's (#1043) result. $\Pr\{|X_1| < c_1, \ldots, |X_k| < c_k\}$ is a non-decreasing function of correlations. Equicorrelated variables. Correlations of form $\lambda_i\lambda_j\rho_{ij}$. Counterexample to show that complete analogue of Slepian's result does not hold in general. Application: confidence region for mean vector.

1027. Sidak, Z. (1971): On probabilities of rectangles in multivariate Student distributions: their dependence on correlations. Ann. Math. Statist., 42, 169-175.

Inequalities for probabilities of rectangles in multivariate case: (i) MVN with general correlation structure, (ii) multivariate t with a special correlation structure.

1028. Siddiqui, M.M. (1967): A bivariate t distribution. Ann. Math. Statist., 38, 162-166. [Correction: 38 (1967), 1594].

Bivariate t distribution derived from BVN. Exact distribution for degrees of freedom n = 1 (bivariate Cauchy) and n = 3. Asymptotic distribution as an application of method of steepest descent.

Silverman, R.W. (1972): See Jenden, D.J., Fairchild, M.D., Mickey, M.R., Silverman, R.W. and Yale, C., #513.

1029. Simonds, J.L. (1963): Application of characteristic vector analysis to photographic and optical response data. J. Opt. Soc. Amer., 53, 968-974.

Analysis of multiresponse curves using principal component analysis. Family of curves specified by characteristic vectors. Example. Graphs and interpretation of results.

1030. Singh, N. (1960): Estimation of parameters of a multivariate normal population from truncated and censored samples. J. Roy. Statist. Soc., B, 22, 307-311.

MLE of means and variances of a k-variate normal distribution based on a sample doubly truncated or censored on s variates. Variates uncorrelated. Information matrix. Numerical example for trivariate case.

Singh, R.P. (1972): See Tracy, D.S. and Singh, R.P., #1132.

1031. Siotani, M. (1959): The extreme value of the generalized distances of the individual points in the multivariate normal sample. Ann. Inst. Statist. Math., 10, 183-208. [Correction: 11 (1960), 220].

Approximate 100α percent points for the statistics $\chi^2_{max} \equiv \max_{\alpha}\{(x_\alpha - x_0)'\Lambda^{-1}(x_\alpha - x_0)\}$, $T^2_{max} \equiv \max_{\alpha}\{(x_\alpha - x_0)'L^{-1}(x_\alpha - x_0)\}$ where Λ is the variance matrix and L its unbiased estimator. Here x_0 is a given point. The ranges of distances between all the points are also studied: $R^2_{max} \equiv \max_{\alpha<\beta}\{(x_\alpha - x_\beta)'\Lambda^{-1}(x_\alpha - x_\beta)\}$ and $S^2_{max} \equiv \max_{\alpha<\beta}\{(x_\alpha - x_\beta)'L^{-1}(x_\alpha - x_\beta)\}$. Bivariate χ^2 and approximation to upper percentage points in the distribution of the maxima for the MVN case. Case of $x_0 = \mu$ and its estimation. Tables.

1032. Siotani, M. (1960): Notes on multivariate confidence bounds. Ann. Inst. Statist. Math., 11, 167-182.

Simultaneous confidence bounds on (i) specific sets of independent comparisons among the k means, (ii) comparison of the other (k-1) means with one, (iii) $\alpha'(\mu_i - \mu_j)$ for all α, $i \neq j$. Procedure to evaluate upper 100α percent points. Tables.

1033. Siotani, M. (1960): A note on the interval estimation related to the regression matrix. Ann. Inst. Statist. Math., 12, 147-149.

Correction of a result due to Roy or SCB for $d_1'Bd_2$. His bounds have a width that does not depend upon d_1, d_2. The new bounds not only correct this discrepancy but are also shown to be narrower.

1034. Siotani, M. (1961): The extreme value of generalized distances and its applications. Bull. Inst. Internat. Statist., 38, 591-599.

Test of hypothesis of slippage i.e., H_0: $\mu_1 = \ldots = \mu_k$ against H_i: $\mu_1 = \ldots = \mu_{i-1} = \mu_i - \delta\mu = \mu_{i+1} = \ldots = \mu_k$ for k populations. Test statistic $T^2_{max} \equiv \max_i[(\bar{X}_i - \bar{X})'L^{-1}(\bar{X}_i - \bar{X})]$, L being estimated variance matrix. SCB for $a'(\mu_i - \bar{\mu})$ for all a, $\bar{\mu}$ being mean of $\mu_1, \ldots, \mu_k$. MV range and SCB on $a'(\mu_i - \mu_j)$ for all a, $i \neq j$. Comparison of k populations with a standard and the problem of SCB on $a'(\mu_i - \mu_0)$.

1035. Siotani, M. (1964): Tolerance regions for a multivariate normal population. Ann. Inst. Statist. Math., 16, 135-153.

Tolerance regions for MVN when (i) Σ known, μ unknown, (ii) μ known, Σ unknown, (iii) both unknown. Simultaneous tolerance regions along the lines of SCB. Approximation to the distribution of positive definite quadratic forms.

1036. Siotani, M. (1964): Interval estimation for linear combinations of means. J. Amer. Statist. Assoc., 59, 1141-1164.

Simultaneous interval estimation. Correlated t's and maxima. Approximation to the probability distribution of the maximum. Principle of inclusion-exclusion. Randomized block design and normal linear regression. $P\{t_i^2 > t^2, t_j^2 > t^2\}$ and table of this probability.

1037. Siotani, M. (1971): An asymptotic expansion of the non-null distribution of Hotelling's generalized T_0^2-statistic. Ann. Math. Statist., 42, 560-571.

Asymptotic expansion for non-null distribution of T_0^2 to terms of $O(n^{-2})$. Non-centrality parameter of a special form. Characteristic function of non-null T_0^2 and its expansion by Welch-James procedure. Inversion. Asymptotic expansion for p.d.f. and c.d.f. Table.

1038. Siotani, M. (1971): Simultaneous confidence intervals relating to the multivariate regression matrices. Tech. Rep. #15, Dept. Statist., Kansas State U.

A simple derivation of the SCI for the double linear combination of elements of a regression matrix (RM). t_{max}^2 and its use. SCI on the contrasts of column vectors of the RM. Comparison of mean vectors of several normal populations. SCI, comparison based on contrasts, SCI on $(\mu_{ir}-\mu_{is})$ for all i and $r \neq s$. Comparison of k regression matrices. SCI for $k = 2$, double linear combination of one linear function of k matrices, finite number of such functions, all linear combinations of the RM's. Percentage points.

1039. Siotani, M. (1971): Asymptotic joint distribution of $\binom{p}{t}$ multiple correlation coefficients between a certain variate and t variates among p other variates (t < p). Tech. Rep. #16, Dept. Statist., Kansas State U.

Covariance between two multiple correlations. Covariance between the elements of a covariance matrix, the determinants of submatrices. Joint asymptotic distributions of the $\binom{p}{t}$ multiple correlations between x_0 and all the tth order subsets of the vector $(x_1,\ldots,x_p)$. Method of differentials.

1040. Siotani, M., Chou, C. and Geng, S. (1971): Asymptotic joint distributions of vector correlation coefficients and of vector alienation coefficients. Tech. Rep. #22, Dept. Statist., Kansas State U.

The joint distribution of the correlation coefficients between $(x_1,\ldots,x_p)$ and the $\binom{p}{t}$ subsets of the vector $(y_1,\ldots,y_q)$ of size t. This is seen immediately to be of the normal form. Paper devoted to determining the covariance matrix between the elements of the vector. Method of differentials. Distribution of the largest squared correlation.

Siotani, M. (1972): See Chou, C. and Siotani, M., #166.

Siotani, M. (1972): See Chou, C. and Siotani, M., #167.

1041. Siskind, V. (1972): Second moments of inverse Wishart-matrix elements. Biometrika, 59, 690-691.

A generalization of a result of Das Gupta. Proof of the result on the expected value of $\{A^{-1}tt'A^{-1}\}$ where A is Wishart and t is a constant vector.

1042. Sitgreaves, R. (1961): Some results on the distribution of the W-classification statistic. Stud. Item. Anal. Pred., (Solomon, ed.), 241-251.

Classification into one of two populations when the mean vectors and common covariance matrix are unknown. Distribution of W-classification statistic: series representation for $n_1 = n_2$. Probabilities of misclassification.

Sitgreaves, R. (1961): See Bowker, A.H. and Sitgreaves, R., #121.

Sitgreaves, R. (1961): See Teichroew, D. and Sitgreaves, R., #1123.

1043. Slepian, D. (1962): The one-sided barrier problem for Gaussian noise. Bell System Tech. J., 41, 463-501.

Probability that a multivariate random variable will be non-negative in the general case. Non-decreasing function of correlation coefficient. Results on stochastic behaviour over time.

1044. Smith, E.H. and Stone, D.E. (1967): A note on the expected coverage of one circle by another, the case of offset aim. SIAM Rev., 3, 51-53.

Distribution of aiming errors: circular normal distribution. Expected coverage of the target circle by the weapon radius circle. Fourier-Bessel integral.

1045. Smith, H., Gnanadesikan, R. and Hughes, J.B. (1962): Multivariate analysis of variance (MANOVA). Biometrics, 18, 22-41.

Analysis of multiresponse data using MANOVA techniques. Tests of significance: (i) Roy's largest root criterion, (ii) Wilks' likelihood ratio criterion, (iii) Hotelling-Lawley sum of roots criterion. Confidence interval estimation using Roy's largest root. MANOVA in the presence of covariables. Techniques illustrated by biochemical example. Interpretation of results.

1046. Smith, W.B. (1968): Bivariate samples with missing values II. Technometrics, 10, 867-868.

Estimation of mean vector and covariance matrix in BVN when some observations are missing. Special case ($p = 2$) of Hocking and Smith's general multivariate method.

Smith, W.B. (1968): See Hocking, R.R. and Smith, W.B., #468.

1047. Smith, W.B. and Pfaffenberger, R.C. (1970): Selection index estimation from partial multivariate normal data. Biometrics, 26, 625-639.

Composite selection index $I_j = b'x_j$. Estimation of b using full and incomplete data records. Uses techniques of Hocking and Smith (#468). Monte Carlo studies. Comparison of estimates based on (i) combination of complete and partial data, (ii) only complete data.

1048. Smith, W.B. and Hocking, R.R. (1972): Wishart variate generator. Appl. Statist., 21, 341-345.

Generation of matrix of sample variances and covariances for sample of size n from MVN (μ, Σ). Bartlett's decomposition. Computer algorithm. Standard FORTRAN.

1049. Smouse, P.E. (1972): The canonical analysis of multiple species hybridization. Biometrics, 28, 361-371.

Multiple discriminant analysis. Classification into one of k groups. Graphical aids. Geometric interpretation. Illustrated by multiple species hybridization.

1050. Snee, R.D. and Andrews, H.P. (1971): Statistical design and analysis of shape studies. Appl. Statist., 20, 250-258.

Characterizing the geometrical appearance of two-dimensional objects. Use of p diameter-to-length ratios. ANOVA vs. principal component analysis of multiple response (or repeated measurements) data.

1051. Snee, R.D. (1972): On the analysis of response curve data. Technometrics, 14, 47-62.

Analysis of repeated measurements experiments using a combination of ANOVA and principal component analysis. Technique more sensitive to differences in the shape of the curve than ANOVA alone of analysis of regression coefficients. Examples to illustrate techniques. Interpretation of results.

1052. Solomon, H. (1961): Classification procedures based on dichotomous response vectors. Stud. Item. Anal. Pred., (Solomon, ed.), 177-186.

Joint distribution of responses of n dichotomous items. Bahadur's representation. LR procedures. Probability of misclassification: risk curves. Comparison of procedures based on dichotomous response vectors with Fisher's discriminant function.

1053. Sondhi, M.M. (1961): A note on the quadrivariate normal integral. Biometrika, 48, 201-203.

Approximation for orthant probability for four dimensional equicorrelated MVN. Probability accurate to 5 decimal places for $\frac{1}{2} \leq \rho \leq 1$.

1054. Sorum, M.J. (1971): Estimating the conditional probability of misclassification. Technometrics, 13, 333-342.

Estimation of the probability of misclassification. Two group classification problem in MVN case with unknown μ and known Σ. Possible estimators: (i) reclassification estimator, (ii) test sample estimators, (iii) estimators based on Okamoto's expansion, (iv) conditional (given sample means) UMVU estimators. Monte Carlo studies. Comparison of estimators on the basis of asymptotic conditional mean square error.

1055. Sorum, M.J. (1972): Three probabilities of misclassification. Technometrics, 14, 309-316.

Classification procedures for two MVN distributions with Σ known and unknown. Probabilities of misclassification: conditional, average, optimum. Distance between probabilities. Distance between a probability and its estimator.

1056. Sowden, R.R. and Ashford, J.R. (1969): Computation of the bivariate normal integral. Appl. Statist., 18, 169-180.

Evaluation of BVN integral. Reduction of two-fold integral to forms involving single integral. Comparison of three methods: Owen's method, Hermite-Gauss quadrature, Simpson's rule. Suitableness of approximation determined on basis of correlation coefficient.

1057. Sprent, P. (1968): Linear relationships in growth and size studies. Biometrics, 24, 639-656.

Analysis of growth data using linear functional relationships. Considers $\Pi y_i^{\alpha_i} = \gamma$, linearized to $\Sigma\alpha_i \log y_i - \log \gamma$. Bivariate and multivariate case with known and unknown Σ. Tests of linearity based on characteristic vectors and eigenvalues. Multivariate generalization of allometry.

1058. Sprent, P. (1972): The mathematics of size and shape. Biometrics, 28, 23-37.

Bivariate and multivariate allometry. Tests of isometry. Linear functional relationships. Shape vectors and size variables (Mosimann) associated with p distances between specified points. Use and interpretation of principal component and factor analyses. Canonical analysis. Discriminant analysis. Work of Mosimann, Jolicoeur and Hopkins compared and contrasted. Alternatives to their techniques.

1059. Spurrell, D.J. (1963): Some metallurgical applications of principal components. Appl. Statist., 12, 180-188.

Combination of techniques of multiple regression and principal component analyses. Principal components of correlation matrix of X's. Regression of each principal component of Y and determination of reduction of variance. Examples and interpretation.

1060. Srivastava, A.B.L. (1960): The distribution of regression coefficients in samples from bivariate non-normal populations I. Theoretical investigation. Biometrika, 47, 61-68.

Edgeworth type bivariate distributions. Distribution of regression coefficient: Mean and variance. Test of significance for $\beta_{21} = 0$. *Effect of non-normality on this test.*

1061. Srivastava, J.N. (1964): On the monotonicity property of the three main tests for multivariate analysis of variance. J. Roy. Statist. Soc., B, 26, 77-81.

Direct proofs of the monotonicity of the power function of each root separately. The criteria are: trace criterion, LR test, Roy's largest root. An error in Roy-Mikhail paper is corrected.

Srivastava, J.N. (1964): See Roy, S.N. and Srivastava, J.N., #979.

1062. Srivastava, J.N. (1965): A multivariate extension of the Gauss-Markov theorem. Ann. Inst. Statist. Math., 16, 63-66.

BLUE for MV populations (not necessarily normal). Estimation of a linear function of all the unknown parameters in the linear model.

1063. Srivastava, J.N. (1966): Some generalizations of multivariate analysis of variance. Multivariate Anal.-I, [Proc. 1st Internat. Symp., Krishnaiah, ed.], 129-145.

A general discussion of MANOVA and its generalizations and applications to (i) multiresponse block designs in which the blocks for different variables are not identical, (ii) ordered responses, resulting in hierarchical designs, (iii) general incomplete MR designs with some responses missing, (iv) response surfaces in MV case. Inapplicability of MANOVA is pointed out in some cases. Examples.

1064. Srivastava, J.N. (1966): On testing hypotheses regarding a class of covariance structures. Psychometrika, 31, 147-164.

Reducible structure. Algebra of reducible patterns. NSC for Σ to be reducible when it is patterned. LR test for the variance matrix to be of the form (patterned) $\alpha_1\Sigma_1+\alpha_2\Sigma_2+\ldots+\alpha_q\Sigma_q$, with Σ_i's being known. Maximization of the likelihood. Tests based on the diagonalizing matrix P. Step-down procedures.

1065. Srivastava, J.N. (1966): Incomplete multiresponse designs. Sankhya, A, 28, 377-388.

General theory of incomplete MR designs is presented via an example. The basic structure is obtained when not all the variates (responses) are measured on each experimental unit. The mathematical formulae relevant to this situation are presented. Procedure for analysis is summarized.

1066. Srivastava, J.N. (1967): On the extension of Gauss-Markov theorem to complex multivariate linear models. Ann. Inst. Statist. Math., 19, 417-437.

Necessary and sufficient conditions are derived for the existence of unique BLUE for linear functions of location parameters in a number of MV linear models: general incomplete multivariate model, hierarchical MV model, multiple design, MV linear model. The Gauss-Markov theorem is generalized.

1067. Srivastava, J.N. (1968): On a general class of designs for multiresponse experiments. Ann. Math. Statist., 39, 1825-1843.

General discussion of incomplete multiresponse designs. Homogeneous and regular incomplete MRD. Transformation of such designs to facilitate MANOVA. Existence of a linear transformation. Removing the singularity of the matrix LL'. Example.

1068. Srivastava, J.N. (1969): Some studies on intersection tests in multivariate analysis of variance. Multivariate Anal.-II, [Proc. 2nd Internat. Symp., Krishnaiah, ed.], 145-170.

Union-Intersection principle. Decomposition of H_0 into meaningful subhypotheses. Problem of optimal choice of sizes of acceptance regions for subhypotheses to maximize the power for H_0. Behaviour of the F-tests. Cases of independent, quasi-independent and dependent decompositions. SANOVA tests. Intersection tests as decompositions of various linear hypotheses.

1069. Srivastava, J.N. and McDonald, L.L. (1969): On the costwise optimality of hierarchical multiresponse randomized block designs under the trace criterion. Ann. Inst. Statist. Math., 21, 507-514.

General incomplete multiresponse designs and the hierarchical multiresponse designs. Under the cost restriction (equal) HMD is superior to GIMD in the sense of having a smaller trace of the covariance matrix of the estimates (of the parameters). The inadvisability of the standard multiresponse design is pointed for some situations. A non-linear programming problem is stated and solved.

1070. Srivastava, J.N. and McDonad, L.L. (1970): On the extension of Gauss-Markov theorem to subsets of the parameter space under complex multivariate linear models. ARL 70-0047, Wright Patterson AFB, Ohio.

Extension of #1066 to complete multiresponse designs. NSC for the unique BLUE to exist for all elements in a subset of the set of all estimable linear functions of the location parameters. Designs considered: general incomplete multiresponse designs and the multiple multiresponse designs.

1071. Srivastava, J.N. and McDonald, L.L. (1971): Analysis of growth curves under the hierarchical models and spline regression, I. ARL 71-0023, Wright Patterson AFB, Ohio.

Potthoff-Roy model and its modiviation: Experimental units S split into subsets such that S_i has observations on $p_1+p_2+\ldots+p_i$ time points and S_{i+1} has in addition to these, observations on p_{i+1} points. Spline regression technique is advocated. Tests of hypotheses under this growth curve model (called hierarchical) are developed by Roy's stepdown procedure. Distribution of the statistics examined. Relationship to standard multiresponse model.

1072. Srivastava, J.N. and McDonald, L.L. (1971): On the costwise optimality of certain hierarchical and standard multiresponse models under the determinant criterion. J. Multivariate Anal., 1, 118-128.

A study similar to that of #1069 with the criterion being the determinant instead of trace. Consideration restricted to randomized block designs.

1073. Srivastava, J.N. and Zaatar, M.K. (1972): On the maximum likelihood classification rule for incomplete multivariate samples and its admissibility. J. Multivariate Anal., 2, 115-126.

Classification of multiresponse observations into one of two p-variate normal populations with unknown mean and known variance. Incomplete MR samples. ML classification rule and its admissibility. Unique Bayes rule and the ML procedure.

1074. Srivastava, M.S. (1965): On the complex Wishart distribution. Ann. Math. Statist., 36, 313-315.

Linear transformations and the Jacobians of the transformations. Distribution of $Y\bar{Y}'$. Wishart and related distributions.

1075. Srivastava, M.S. (1965): Some tests for the intraclass correlation model. Ann. Math. Statist., 36, 1802-1806.

Tests of the hypothesis when $\Sigma_i = \sigma_i^2[(1-\rho_i)J+\rho_i I]$: (i) H_0: $\rho_i = \rho_j$, $\sigma_i = \sigma_j$ $(i \neq j)$, (ii) H_0: $\rho_i = 0$, (iii) H_0: $\rho_i = \rho_j$, $(i \neq j)$, against appropriate alternatives for k populations. LR and Roy's UI.

1076. Srivastava, M.S. (1966): On a multivariate slippage problem I. Ann. Inst. Statist. Math., 18, 299-305.

Slippage in the variance matrices. Admissible procedures. Bayes technique. Simple loss function (zero-one). Kiefer-Schwartz prior. MV analogue of Cochran's statistic.

1077. Srivastava, M.S. (1967): Comparing distances between multivariate populations - the problem of minimum distance. Ann. Math. Statist., 38, 550-556.

Classificatory problem for Mahalanobis distance point of view. Nearest distance: Bayes procedure for symmetric prior. Optimal decision rule for simple loss function. Admissibility. ML rule and its admissibility under simple loss function. Σ known and unknown.

1078. Srivastava, M.S. (1967): Classification into multivariate normal populations when the population means are linearly restricted. Ann. Inst. Statist. Math., 19, 473-478.

Tests of hypothesis on the means. Classificatory and discriminatory procedures. ML rules for classifying a sample as belonging to one of k MVN populations. Cases of known and unknown variances assumed equal. Admissibility and (a.e.) unique Bayes nature of ML rule. Simple loss function.

1079. Srivastava, M.S. (1968): On the distribution of a multiple correlation matrix: noncentral multivariate beta distributions. Ann. Math. Statist., 39, 227-232. [Correction: 39 (1968), 1359].

Non-null distribution of multiple correlation matrix when (i) one of two sets of variables is fixed (MANOVA) and (ii) both sets are random (canonical correlations). Zonal polynomials. Transformations and Jacobians. Noncentral multivariate β-distributions.

Srivastava, M.S. (1971): See Khatri, C.G. and Srivastava, M.S., #619.

1080. Srivastava, S.R. and Bancroft, T.A. (1967): Inferences concerning a population correlation coefficient from one or possibly two samples subsequent to a preliminary test of significance. J. Roy. Statist. Soc., B, 29, 282-291.

Simple correlation in the two sample BVN case. Estimation and tests of ρ based on preliminary tests of significance. Asymptotic distribution theory based on logarithmic transformation of r. MSE studies. Power and size of tests.

1081. Srivastava, S.R. (1969): A note on the estimation of a correlation coefficient incorporating a preliminary test of significance. Trabajos Estadist., 20, 113-115.

Continuation of #1080. Comparison of two estimators for ρ based on presence or absence of prior information concerning $\rho_1 = \rho_2$. MSE of estimators.

1082. Srivastava, V.K. (1970): On the expectation of the inverse of a matrix. Sankhya, A, 32, 236.

Proof of $E(Q^{-1}) \geq \{E(Q)\}^{-1}$. Alternative proof for result of Groves and Rothenberg.

1083. Steck, G.P. (1958): A table for computing trivariate normal probabilities. Ann. Math. Statist., 29, 780-800. [Correction: 30 (1959), 1297].

Evaluation of trivariate normal integral in terms of G(x) (the distribution function of N(0,1) variable), T(h, a) (a function of the univariate normal) and a function S(h, a, b). Extensive tables of S(h, a, b). Seven-point Gaussian quadrature formula.

1084. Steck, G.P. (1962): Orthant probabilities for the equicorrelated multivariate normal distribution. Biometrika, 49, 433-445.

Review of results pertaining to orthant probabilities in the equicorrelated case. Formula relating orthant probabilities for positive and negative ρ. Approximation for $p = 4$ and $\rho \geq 0$. Gram-Charlier Type A series approximation: useful for higher dimensions when ρ is near ± 1. Tables.

Steck, G.P. (1962): See Owen, D.B. and Steck, G.P., #860.

1085. Steck, G.P. and Owen, D.B. (1962): A note on the equicorrelated multivariate normal distribution. Biometrika, 49, 269-271.

Relationships in equicorrelated MVN which are useful for determining $Pr(X_1 \leq t, X_2 \leq t, \ldots, X_p \leq t)$.

Steck, G.P. (1964): See Gupta, S.S., Pillai, K.C.S. and Steck, G.P., #421.

1086. Steffens, F.E. (1969): Critical values for bivariate Student t-tests. J. Amer. Statist. Assoc., 64, 637-646.

Tests of $\mu_1 = 0$ and $\mu_2 = 0$ based on bivariate t distribution. Critical points of bivariate t for several alternatives: (i) $\mu_1 \neq 0$ or $\mu_2 \neq 0$, (ii) $\mu_1 > 0$ or $\mu_2 \neq 0$, (iii) $\mu_1 > 0$ or $\mu_2 > 0$, (iv) $\mu_1 > 0$ and $\mu_2 > 0$, (v) $\mu_1 \neq 0$ and $\mu_2 \neq 0$, (vi) $\mu_2 \neq 0$. Tables. Asymptotic approximation using Student's t. Tables of adequacy of approximation.

1087. Steffens, F.E. (1969): A stepwise multivariate t-distribution. South African Statist. J., 3, 17-26.

Definition of a stepwise multivariate t-distribution (central and noncentral). Inclusion or exclusion of variables (in a fixed order) in stepwise regression analysis. Procedure for testing the equality of slopes of two simple linear regression lines with subsequent fitting of parallel lines. Testing equality of intercepts.

1088. Steffens, F.E. (1970): Power of bivariate Studentized maximum and minimum modulus tests. J. Amer. Statist. Assoc., 65, 1639-1644.

Tests of $\mu_1 = \mu_2 = 0$ based on bivariate t distribution. Study of power function for (i) $\mu_1 \neq 0$ or $\mu_2 \neq 0$ (MAXMOD) and (ii) $\mu_1 \neq 0$ and $\mu_2 \neq 0$ (MINMOD). Power tabulated for $\alpha = 0.05$, d.f. = 1, 2, 5, 10, 20, 50, ∞ and $\delta_1, \delta_2 = 0$ (1)5. Asymptotic approximations for power functions. Tabulation of marginal power of MAXMOD test.

Stein, C.M. (1961): See James, W. and Stein, C.M., #510.

1089. Stein, C.M. (1962): Confidence sets for the mean of a multivariate normal distribution. J. Roy. Statist. Soc., B, 24, 265-296.

Estimation of the mean vector of MVN when Σ is known. Point estimation. Sample mean is not best estimator when the loss is a nonsingular quadratic function of the error vector. Confidence sets. Geometrical size and shape. Probability of covering false values. Bayesian sets. Discussion of the paper.

Stein, C.M. (1963): See Giri, N., Kiefer, J. and Stein, C.M., #359.

1090. Stein, C.M. (1966): An approach to the recovery of inter-block information in balanced incomplete block designs. Res. Papers Statist. Festschr. Neyman, (David, ed.), 351-366.

James-Stein estimator of mean vector. Recovery of inter-block information as an application of the estimate of the mean of a MVN. Designs admitting a completely orthogonal analysis of variance. Intra-block estimate of the vector of treatment contrasts. Reduction in expected value of error sum of squares if dimension ≥ 3. Possibility of combining techniques with those of Yates.

Steinberg, L. (1963): See Krishnaiah, P.R., Hagis, P. Jr. and Steinberg, L., #641.

Steinberg, L. (1965): See Krishnaiah, P.R., Hagis, P. Jr. and Steinberg, L., #648.

1091. Steyn, H.S. and Roux, J.J.J. (1970): A method for obtaining generating functions of conditional distributions and some applications. South African Statist. J., 4, 59-66.

Probability generating functions and moment generating functions for conditional distributions. Application in multivariate analysis: MVN and Wishart distributions.

1092. Steyn, H.S. and Roux, J.J.J. (1972): Approximations for the noncentral multivariate Wishart distribution. South African Statist. J., 6, 165-173.

Approximations to noncentral Wishart. Differential equations for the moment generating function of the noncentral distribution. Moments. Approximation of the noncentral distribution by a central distribution using differential equation method. Use of generalized Laguerre polynomials for approximations to noncentral distributions.

Stoller, D.S. (1962): See Scheuer, E.M. and Stoller, D.S., #1005.

Stone, D.E. (1961): See Smith, E.H. and Stone, D.E., #1044.

1093. Stone, M. (1964): Comments on a posterior distribution of Geisser and Cornfield. J. Roy. Statist. Soc., B, 26, 274-276.

Failure of the non-integrable density used by Geisser and Cornfield as a prior density, to be justified in the asymptotic sense to be the limit of integrable priors. The Fisher-Cornish density cannot be viewed as the 'probability limit' for any sequence of densities. Bayesian inferences in MV analysis.

1094. Strawderman, W.E. (1971): Proper Bayes minimax estimators of the multivariate normal mean. Ann. Math. Statist., 42, 385-388.

Proper Bayes minimax estimation of MVN mean. Existence for $p \geq 5$. Proper prior distributions for $p \geq 5$. A class of minimax estimators.

1095. Strawderman, W.E. and Cohen, A. (1971): Admissibility of estimators of the mean vector of a multivariate normal distribution with quadratic loss. Ann. Math. Statist., 42, 270-296.

Spherically symmetric (SS) MV case. Generalized Bayes (GB) estimators. Relation between symmetry of prior distribution, convolutions with normal density and symmetry on GB. NSC for GB. NSC for bounded risk, spherically symmetric estimator to be admissible. Proper Bayes estimators and their characterization. Bounded and unbounded risk.

1096. Strawderman, W.E. (1972): On the existence of proper Bayes minimax estimators of the mean of a multivariate normal distribution. Proc. Sixth Berkeley Symp. Math. Statist. Prob., 1, 51-55.

Extension of #1094. Non-existence of spherically symmetric proper Bayes minimax estimators for p = 3, 4. Bounds on the bias of a minimax estimator. Quadratic loss function. MVN with I as the covariance matrix.

1097. Stuart, A. (1958): Equally correlated variates and the multinormal integral. J. Roy. Statist. Soc., B, 20, 373-378.

Generating MV (correlated) random variables from an uncorrelated set. Normal case. Application to (i) t test of equality of means, (ii) evaluation of P{X > 0} and a generalization of it, (iii) approximation of sampling distributions under randomization.

Studden, W.J. (1970): See Gupta, S.S. and Studden, W.J., #425.

1098. Subrahmaniam, Kathleen (1971): Discrimination in the presence of covariables. South African Statist. J., 5, 5-14.

A unified approach to the theory of discriminant analysis in the presence of covariables by partitioning the D^2-statistic. It is demonstrated that Rao's and Cochran-Bliss methods are equivalent. Distributions of test statistic in the null and non null cases for various hypotheses.

1099. Subrahmaniam, Kathleen and Subrahmaniam, Kocherlakota (1971): On the distribution of $(D^2_{p+q}-D^2_q)$ statistic: Percentage points and the power of the test. Tech. Rep. #7, Dept. Statist., U. Manitoba.

The distribution of $(D^2_{p+q}-D^2_q)$ statistic: null and non-null cases. Comparison of the procedure with the conditional test of Cochran and Bliss. Asymptotic expressions for the p.d.f. of the unconditional distribution of $D^2_{p+q}-D^2_q$. Study of the power function behaviour. Tables of 10, 5 and 1 percent points of the Rao test. Power function tabled for selected values of N, p and q. Graphs.

1100. Subrahmaniam, Kathleen and Subrahmaniam, Kocherlakota (1972): On the confidence region comparison of some solutions for the multivariate Behrens-Fisher problem. Tech. Rep. #30, Dept. Statist., U. Manitoba.

Comparison of the approximate degrees of freedom solution Bennett's solution and James' series solution to Behrens-Fisher problem. Volumes of the ellipsoids of the confidence regions under various procedures. Monte Carlo simulation of Wishart distribution. Tables of the relative efficiency measured by ratio of the volumes.

Subrahmaniam, Kathleen (1972): See Subrahmaniam, Kocherlakota and Subrahmaniam, Kathleen, #1103.

1101. Subrahmaniam, Kocherlakota (1970): On some applications of Mellin transforms to statistics: Dependent random variables. SIAM J. Appl. Math., 19, 658-662.

Mellin transforms defined in the bivariate case. Case of negative random variables. Inversion of the transforms. Applications to order statistics and the distribution of the ratio X/Y when (X, Y) is standardized BVN with correlation ρ.

Subrahmaniam, Kocherlakota (1971): See Subrahmaniam, Kathleen and Subrahmaniam, Kocherlakota, #1099.

1102. Subrahmaniam, Kocherlakota (1972): On some functions of matrix argument. Tech. Rep. #26, Dept. Statist., U. Manitoba

Some properties of generalized hypergeometric function and Laguerre polynomials of matrix argument. Integral transforms, generating functions, partial differential equations. Quadratic transformation and its non-existence by direct generalization to matrix form. A conjecture. Saalschützian identity.

Subrahmaniam, Kocherlakota (1972): See Subrahmaniam, Kathleen and Subrahmaniam, Kocherlakota, #1100.

1103. Subrahmaniam, Kocherlakota and Subrahmaniam, Kathleen (1972): On the multivariate Behrens-Fisher problem. Tech. Rep. #29, Dept. Statist., U. Manitoba.

Related to #1100. Comparison of the solutions in terms of the level of significance and the power. A representation for the power of the tests. Monte Carlo procedures. Tables.

Sudjana (1972): See Pillai, K.C.S. and Sudjana, #906.

1104. Sugiura, N. and Nagao, H. (1968): Unbiasedness of some test criteria for the equality of one or two covariance matrices. Ann. Math. Statist., 39, 1686-1692.

LR tests on the covariance matrices: (i) H_0: $\Sigma = \Sigma_0$, (ii) H_0: $\Sigma = \Sigma_0$, $\mu = \mu_0$, (iii) $\Sigma_1 = \Sigma_2$, (iv) $\Sigma = \sigma^2 I$. Generalization to k sample case. Unbiasedness of the tests, i.e., $P_H(\omega_1) \geq P_A(\omega_1)$ when H and A are the null and the alternative respectively. Also, ω_1 is the critical region.

1105. Sugiura, N. (1969): Asymptotic expansions of the distributions of the likelihood ratio criteria for covariance matrix. Ann. Math. Statist., 40, 2051-2063.

Asymptotic distributions of LR criteria. Box's technique. Characteristic functions of modified criteria for (i) $\Sigma = \Sigma_0$, (ii) $\Sigma = \Sigma_0$, $\mu = \mu_0$, (iii) $\Sigma = \sigma^2 I$. Null and non-null distributions. Asymptotic expansion of characteristic function.

1106. Sugiura, N. and Fujikoshi, Y. (1969): Asymptotic expansions of the non-null distributions of the likelihood ratio criteria for multivariate linear hypothesis and independence. Ann. Math. Statist., 40, 942-952.

Asymptotic distribution of LR criteria. Characteristic function and its expansion in non-null case for linear hypotheses and independence. Derivation of hth moment of LR criterion for independence in non-null case. Zonal polynomials and their properties.

1107. Sugiura, N. (1971): Note on some formulas for weighted sums of zonal polynomials. Ann. Math. Statist., 42, 768-772.

Zonal polynomials. Identities involving zonal polynomials expressed in terms of traces of the matrix and its powers.

1108. Sugiura, N. and Nagao, H. (1971): Asymptotic expansion of the distribution of the generalized variance for noncentral Wishart matrix, when $\Omega = O(n)$. Ann. Inst. Statist. Math., 23, 469-475.

Asymptotic expansion for $|S\Sigma^{-1}/n|$ using characteristic function technique. Noncentral case when $\Omega = O(n)$. Generalization of the results of Fujikoshi.

1109. Sugiura, N. (1972): Locally best invariant test for sphericity and the limiting distributions. Ann. Math. Statist., 43, 1312-1316.

New criterion for sphericity test and its properties. Locally best invariant test. Distribution of test statistic. Limiting distribution of the statistic $tr\ S^2/(tr\ S)^2$ under various alternatives. Comparison with LR criterion.

1110. Sugiyama, T. (1965): On the distribution of the latent vectors for principal component analysis. Ann. Math. Statist., 36, 1875-1876.

Exact distribution of latent vectors of the sample covariance matrix for the two dimensional case. Distribution in terms of hypergeometric series.

1111. Sugiyama, T. (1966): On the distribution of the largest latent root and the corresponding latent vector for principal component analysis. Ann. Math. Statist., 37, 995-1001.

Exact distribution of latent vector corresponding to the largest root of a sample covariance matrix when Σ is arbitrary. Exact distribution of largest latent root when $\Sigma = \sigma^2 I$. Application of zonal polynomials.

1112. Sugiyama, T. (1967): On the distribution of the largest latent roots of the covariance matrix. Ann. Math. Statist., 38, 1148-1151.

Continuation of #1111. Arbitrary Σ. The distribution of the largest root in zonal polynomials for the null case.

1113. Sugiyama, T. (1967): Distribution of the largest latent root and the smallest latent root of the generalized B statistic and F-statistic in multivariate analysis. Ann. Math. Statist., 38, 1152-1159.

The distribution of the largest and smallest roots, in the null case, of the MV beta and F variates, i.e., the roots of $|U_1-(U_1+U_2)\lambda| = 0$, $|U_1-fU_2| = 0$. Distribution functions expressed in terms of zonal polynomials. Integrals involving such polynomials.

Sugiyama, T. (1967): See Fukutomi, K. and Sugiyama, T., #320.

Sugiyama, T. (1969): See Pillai, K.C.S. and Sugiyama, T., #894.

1114. Sugiyama, T. (1970): Joint distribution of the extreme roots of a covariance matrix. Ann. Math. Statist., 41, 655-657.

Distribution of the roots (e_1, e_p) and of the ratio e_1/e_p in zonal polynomials for the null case. Application to sphericity test.

1115. Sugiyama, T. (1972): Approximation for the distribution of the largest latent root of a Wishart matrix. Austral. J. Statist., 14, 17-24.

Generalized Laguerre polynomials and infinite series involving such polynomials. Distribution of the largest latent root of central and noncentral Wishart and approximation using χ^2.

Symons, M.J. (1971): See Scott, A.J. and Symons, M.J., #1015.

Symons, M.J. (1971): See Scott, A.J. and Symons, M.J., #1016.

1116. Tallis, G.M. (1961): The moment generating function of the truncated multinormal distribution. J. Roy. Statist. Soc., B, 23, 223-229.

Moment generating function of the truncated MVN with truncation on one side in each coordinate. Expectation of X_i and $X_i X_j$ by differentiation. Application to culling. Standard MVN.

1117. Tallis, G.M. and Young, S.S.Y. (1962): Maximum likelihood estimation of parameters of the normal, lognormal, truncated normal and bivariate normal distributions from grouped data. Austral. J. Statist., 4, 49-54.

Censoring in BVN at several points on the X and Y axes. ML estimation, using method of scoring. Asymptotic goodness-of-fit test.

1118. Tallis, G.M. (1963): Elliptical and radial truncation in normal populations. Ann. Math. Statist., 34, 940-944.

Elliptical truncation in MVN (standard) and the m.g.f. of the truncated distribution. Application to problems of selection.

1119. Tallis, G.M. (1965): Plane truncation in normal populations. J. Roy. Statist. Soc., B, 27, 301-307.

The m.g.f.'s of truncated MVN distributions when truncated by (i) single plane, (ii) $q \leq p$ planes, (iii) group truncation involving selection in different variates. Mean and variance of truncated distributions.

1120. Tan, W.Y. (1969): Note on the multivariate and the generalized multivariate beta distributions. J. Amer. Statist. Assoc., 64, 230-241.

Decomposition of MV (i) Beta type I into Beta type I and Dirichlet variates, (ii) Beta type II into Beta type II and inverted Dirichlet. Submatrices and their distributions.

1121. Tan, W.Y. and Guttman, I. (1971): A disguised Wishart variable and a related theorem. J. Roy. Statist. Soc., B, 33, 147-152.

Jacobians of some transformations. Distribution of $R = P'VP$ when PP' is Wishart and V is constant.

Tarter, M.E. (1969): See Kowalski, C.J. and Tarter, M.E., #636.

Tate, R.F. (1961): See Olkin, I. and Tate, R.F., #852.

Tate, R.F. (1965): See Hannan, J.F. and Tate, R.F., #443.

1122. Tate, R.F. (1966): Conditional normal regression models. J. Amer. Statist. Assoc., 61, 477-489.

Sample correlation coefficient when the underlying distribution has discrete and continuous component. Conditional distribution of $r^2/(1-r^2)$. Limiting distribution. Tests of hypotheses on ρ: (i) $\rho = 0$, (ii) $\rho = \rho_0$. Asymptotic tests.

Tate, R.F. (1966): See Prince, B.M. and Tate, R.F., #916.

1123. Teichroew, D. and Sitgreaves, R. (1961): Computation of an empirical sampling distribution for the W-classification statistic. Stud. Item. Anal. Pred., (Solomon, ed.), 252-275.

Determination of cumulative distribution function of W. Monte Carlo techniques. Generating random Wishart matrices. Generation of normal and multivariate normal variates. Approximating distributions using Chebychev polynomials.

Terragno, P.J. (1964): See Guenther, W.C. and Terragno, P.J., #408.

1124. Thompson, W.A. Jr. (1962): Estimation of dispersion parameters. J. Res. Nat. Bur. Standards, Sect. B, 66, 161-164.

Simultaneous confidence bounds for roots of a Wishart matrix. Tables of percentage points of smallest and largest roots of a bivariate Wishart matrix. Application to precision of instruments.

Thompson, W.A. Jr. (1968): See Hanumara, R.C. and Thompson, W.A. Jr., #444.

Thompson, W.A. Jr. (1972): See Brindley, E.C. Jr. and Thompson, W.A. Jr., #124.

1125. Tiao, G.C. and Zellner, A. (1964): On the Bayesian estimation of multivariate regression. J. Roy. Statist. Soc., B, 26, 277-285.

Multivariate linear model with equal X matrices. Bayesian estimation. Posterior distribution of the regression matrix and covariance matrix. Posterior (marginal) distribution of Σ. Inverted Wishart distribution. Posterior (marginal) of first component of β. Joint distribution of components of β. Multivariate t distribution. Decomposition theorem for distribution of β.

Tiao, G.C. (1968): See Box, G.E.P. and Tiao, G.C., #122.

1126. Tiao, G.C. and Fienberg, S. (1969): Bayesian estimation of latent roots and vectors with special reference to the bivariate normal distribution. Biometrika, 56, 97-108.

Latent roots and vectors of Wishart matrix in BVN. Bayesian approach. Joint posterior for θ, the angle of the canonical transformation and V the ratio of the larger root to the sum of the roots. Properties of the joint distribution and the marginals. Exact and approximate distributions. Posterior distribution of the larger root. Inverted χ^2 approximation and scaled inverted χ^2 approximation to this distribution. Scaled inverted χ^2 approximation in general. The prior is the non-information prior.

1127. Tihansky, D.P. (1972): Properties of the bivariate normal cumulative distribution. J. Amer. Statist. Assoc., 67, 903-905.

Equidistributional contours of BVN. Functional relationship between lower limits of the upper c.d.f. Shifts in contours and relationship with correlation. Concavity and correlation.

1128. Tiku, M.L. (1971): A note on the distribution of Hotelling's generalized T_0^2. Biometrika, 58, 237-241.

Hotelling's generalized T_0^2. Chi square and inverted chi square approximations. Four moment approximation. Example.

1129. Timm, N.H. (1970): The estimation of variance-covariance and correlation matrices from incomplete data. Psychometrika, 35, 417-437.

Estimation using four methods (i) principal component solution of known data to estimate unknown, (ii) substitute mean of the unknown, (iii) multiple regression technique, (iv) complete data only. Examination of the procedure based on selected data from literature. Effect of sample size, number of variables, percent missing and average inter-correlation of variables.

Tomsky, J. (1972): See McNolty, F. and Tomsky, J., #786.

1130. Tong, Y.L. (1970): Some probability inequalities of multivariate normal and multivariate t. J. Amer. Statist. Assoc., 65, 1243-1247.

Probabilities of rectangles in multivariate t and MVN. Equicorrelated case. Relationship between probabilities of k-dimensional and m-dimensional rectangles. Moment inequalities of nonnegative random variables. Applications.

1131. Tracy, D.S. and Dwyer, P.S. (1969): Multivariate maxima and minima with matrix derivatives. J. Amer. Statist. Assoc., 64, 1576-1599.

Determination of extrema. Second order derivatives. Extrema under constraints. Table of formulae for some commonly occurring forms. Applications: quadratic forms, general least squares, MVN, MLE in multivariate linear regression.

1132. Tracy, D.S. and Singh, R.P. (1972): A new matrix product and its applications in partitioned matrix differentiation. Statistica Neerlandica, 26, 143-157.

Definition of a new Kronecker product. Applications to differentiation of a function of matrix argument when matrix is partitioned. Examples.

1133. Trawinski, I.M. and Bargmann, R.E. (1964): Maximum likelihood estimation with incomplete multivariate data. Ann. Math. Statist., 35, 647-657.

MLE with missing observations. General linear model with some parts of observation vectors missing by design, not randomly. Estimation of regression matrix and covariance matrix. Iterative procedures. Tests of linear hypothesis of the type $C\xi = 0$. Examples.

1134. Trawinski, I.M. (1967): An algorithm for obtaining the zero of a function of the dispersion matrix in multivariate analysis. J. Amer. Statist. Assoc., 62, 114-123.

MLE of covariance matrix for multivariate experiments in which measurements on certain components of observation vector are intentionally omitted. Algorithm based on Newton-Raphson method. Example of technique.

1135. Troskie, C.G. (1967): Noncentral multivariate Dirichlet distributions. South African Statist. J., 1, 21-32.

Real and complex Wishart distributions. Derived distributions. Real and complex multivariate Dirichlet of Types I and II. Moments of Dirichlet determinant. Application to multivariate tests of hypotheses. Moments of LR criterion.

1136. Troskie, C.G. (1969): The generalized multiple correlation matrix. South African Statist. J., 3, 109-121.

Generalized multiple correlation matrix: distribution and properties. Central and noncentral moments. Tests of hypotheses on the independence of two sets of variates: (i) LR criterion, (ii) Hotelling's canonical correlation, (iii) Pillai's criterion, (iv) Roy's largest root criterion. Moments of LR criterion. Exact distributions of $|R|$ and $|I-R|$ in linear case. Generalized multiple coherence matrix. Relationship of other "correlation coefficients" to generalized multiple correlation matrix.

1137. Troskie, C.G. (1971): The distributions of some test criteria in multivariate analysis. Ann. Math. Statist., 42, 1752-1757.

Tests of independence of sets of variates. Criteria based on multiple correlation matrix: (i) Wilks' Λ, (ii) Pillai's V, (iii) Hotelling's T_0^2, (iv) Roy's largest root. Exact null and non-null distributions of criteria. Zonal polynomials.

1138. Troskie, C.G. (1972): The distributions of some test criteria depending on multivariate Dirichlet distributions. South African Statist. J., 6, 151-163.

Definition of multivariate Dirichlet of types I and II derived from Wishart distributions. Integrals involving matrix arguments and zonal polynomials. Distribution of the sum of multivariate Dirichlet random variables. Distribution of trace of sum and other related statistics.

1139. Troskie, C.G. and Money, A.H. (1972): Some further properties of test criteria in multivariate analysis. South African Statist. J., 6, 121-134.

Tests on regression matrix. Distribution of $X^{(1)}$ given $X^{(2)}$ and of $X^{(2)}$ given $X^{(1)}$. Interpretation of H_0: $B = 0$ as (i) independence, $X^{(1)}$, $X^{(2)}$ both random, (ii) non-estimability of $X^{(1)}$ given $X^{(2)}$, $X^{(2)}$ fixed. Distribution of smallest root r_1^2 and of $Q = tr\{R^{-1}(I-R)\}$, R is the multiple correlation matrix. Moments of criteria. Generating functions.

Truax, D.R. (1966): See Karlin, S. and Truax, D.R., #575.

1140. Truett, J., Cornfield, J. and Kannel, W. (1967): A multivariate analysis of the risk of coronary heart diseases in Framingham. J. Chronic Diseases, 20, 511-524.

Linear discriminant function analysis of two populations. Multivariate logistic function. Estimation of risk. Effects of deviations from multivariate normality. Application to study of risk of heart disease using seven (not necessarily normal) variables.

1141. Tsukibayashi, S. (1962): Estimation of bivariate parameters based on range. Rep. Statist. Appl. Res. JUSE, 9, 10-23.

(X,Y) is BVN with mean zero and variance matrix Σ. $W_{v(n)} = v(n)-v(1)$, the range of v which is equal to X or Y. $W_{y(n)} = y_n-y_1$, where y_i is the value of y associated with X_i. Estimation of Σ using this range statistic. Application to ANCOVAR. Extension to $p > 2$ in the independent case.

1142. Tukey, J. (1962): The future of data analysis. Ann. Math. Statist., 33, 1-67. [Correction: 33 (1962), 812].

Data analytic techniques in general. Applications to multiresponse data: Two samples, factor analysis, classification problem. Bibliographic.

Tukey, J. (1966): See Kurtz, T.E., Link, R.F., Tukey, J. and Wallace, D.L., #696.

1143. Uematu, T. (1959): Note on the numerical computation in the discrimination problem. Ann. Inst. Statist. Math., 10, 131-135.

Solving equation $(ANA')u = \lambda Bu$, *in the linear discriminant function evaluation. Degeneration of the equation and its applications. Computational procedure. Comparison with the method using inversion of the matrix B.*

1144. Uematu, T. (1964): On a multidimensional linear discriminant function. Ann. Inst. Statist. Math., 16, 431-437.

Classification into $s \geq 2$ *populations. Generalization of Hayashi's technique. Use of m (in contrast with one) linear discriminant functions of the form* $y_i = \Sigma a_{ij} x_j$. *Coefficients* $\{a_{ij}\}$ *are determined so as to maximize* $\lambda = 1 - [|\Sigma_\omega|/|\Sigma|]$. *Case for m = 2 discussed in detail.*

Uppuluri, V.R.R. (1970): See Olson, W.H. and Uppuluri, V.R.R., #857.

1145. Urbakh, V. Yu. (1971): Linear discriminant analysis: loss of discriminating power when a variate is omitted. Biometrics, 27, 531-534.

Selection of variates for use in a discriminant analysis. Estimation of the reduction of the discriminating power when a single variate from a complete set of p variates is excluded after formation of a discriminant function. Criterion proposed for omission of a variate without an increase in probability of misclassification. Tables.

1146. Van der Vaart, H.R. (1961): On certain characteristics of the distribution of the latent roots of a symmetric random matrix under general conditions. Ann. Math. Statist., 32, 864-873.

Latent roots of a symmetric matrix F estimate the roots of Σ, *the expected value of F, biasedly. Both median and mean bias are included. Relations between covariances of the latent roots, covariances of F, mean bias of the latent roots. Symmetry conditions.*

1147. Van der Vaart, H.R. (1965): A note on Wilks' internal scatter. Ann. Math. Statist., 36, 1308-1312.

A matrix X of which row one is (1,...,1) and all the columns are independent identical r.v.'s with expectation μ and variance Σ, has the determinant D. $E(D^2) = \{(k+1)!\}\Delta$ where Δ = determinant of Σ. Wilks' theorem on internal scatter and its proof. Unbiased estimation of generalied variance.

Van Eeden, C. (1972): See Kraft, C.H., Olkin, I. and Van Eeden, C., #638.

1148. Van Yzeren, J. (1972): A bivariate distribution with instructive properties as to normality, correlation and dependence. Statistica Neerlandica, 26, 55-56.

Example of the BV distributions where (i) (X, Y) are not jointly normal but each marginal is N(0,1), (ii) ρ = 0, X+Y and X-Y are not normal but (X, Y) are dependent.

Varadarajan, V.S. (1963): See Rao, C.R. and Varadarajan, V.S., #933.

Varady, P.D. (1966): See Dunn, O.J. and Varady, P.D., #269.

1149. Villegas, C. (1969): On the a priori distribution of the covariance matrix. Ann. Math. Statist., 40, 1098-1099.

Fiducial argument in favour of non-informative a priori distribution of the covariance matrix.

1150. Wagle, B. (1968): Multivariate beta distribution and a test for multivariate normality. J. Roy. Statist. Soc., B, 30, 511-516.

Multivariate beta and Wishart distributions. Distribution of the product of the square root of a Wishart and multivariate beta variate: normal. Test for multivariate normality in the presence of nuisance parameters. Multivariate generalization of Durbin's method.

Waikar, V.B. (1969): See Krishnaiah, P.R. and Waikar, V.B., #656.

Waikar, V.B. (1970): See Krishnaiah, P.R. and Waikar, V.B., #660.

Waikar, V.B. (1971): See Krishnaiah, P.R. and Waikar, V.B., #663.

Waikar, V.B. (1971): See Krishnaiah, P.R. and Waikar, V.B., #664.

1151. Waikar, V.B., Chang, T.C. and Krishnaiah, P.R. (1972): Exact distributions of a few arbitrary roots of some complex random matrices. Austral. J. Statist., 14, 84-88.

Joint p.d.f. of any few unordered roots of a class of noncentral random complex matrices. Zonal polynomials. Matrices considered: (i) A matrix of interest to physicists, (ii) Wishart matrix, (iii) MANOVA matrix (iv) canonical correlation matrix.

1152. Walker, M.A. (1967): Some critical comments on "An analysis of crimes by the method of principal components", by B. Ahamad. Appl. Statist., 16, 36-39.

Critcism of the use of principal component analysis in Ahamad's paper. Correction for "time" effect prior to principal component analysis. Interpretation of the regression analysis using principal components is questioned.

Wallace, D.L. (1966): See Kurtz, T.G., Link, R.F., Tukey, J.W. and Wallace, D.L., #696.

1153. Wang, Y.Y. (1972): Selection problems under multivariate normal distribution. Biometrics, 28, 223-233.

Genetic selection when p genetic variates and p environmental variates are considered simultaneously. Selection under p-dimensional MVN. Proportion of genetic deviates and distribution of genetic variates after selection represented in terms of quadratic forms. Problem of k-cycle selection: distribution of genetic variate and its expectation. Evaluation of probability integral for MVN.

1154. Wani, J.K. and Kabe, D.G. (1971): Generalized inverse and its applications in classical normal multivariate regression theory. Trabajos Estadist., 22, 137-146.

Application of the generalized inverse of a matrix to problems in multivariate regression analysis. Model: $Y = BX+E$ with B a matrix and X not necessarily of full rank. Distribution of the estimators of B and Σ. Distribution of the test criterion for testing hypotheses concerning B. Prediction of future observations. Estimation of fixed variates.

Warner, J.L. (1971): See Andrews, D.F., Gnanadesikan, R. and Warner, J.L., #43.

1155. Warner, S.L. (1963): Multivariate regression of dummy variates under normality assumptions. J. Amer. Statist. Assoc., 58, 1054-1063.

Regression analysis when the independent variables are MVN and the dependent variables are 0, 1 variables. $E(Y|X)$ using Bayes Rule. MLE of $E(Y|X)$. Connection with discriminant function analysis.

1156. Watson, G.S. (1964): A note on maximum likelihood. Sankhya, A, 26, 303-304.

MLE of the dispersion matrix in MVN. Applied to situation concerning instrumental variables.

Watson, G.S. (1972): See Gleser, L.J. and Watson, G.S., #377.

1157. Watterson, G.A. (1959): Linear estimation in censored samples from multivariate normal populations. Ann. Math. Statist., 30, 814-824.

Censoring of types A, B, C in BVN and MVN. Estimation of parameters by linear functions of observations. Estimators having reasonably small variances. Tables of efficiencies. Detailed discussion of bivariate case.

1158. Waugh, F.V. and Fox, K.A. (1957): Graphic computation of $R_{1.23}$. J. Amer. Statist. Assoc., 52, 470-481. [Correction: 53 (1958), 1031].

Graphical techniques for the evaluation of the multiple correlation coefficient. $R_{1.23}$ is given in terms of r_{12}, r_{13}, r_{23}.

1159. Webster, J.T. (1970): On the application of the method of Das in evaluating a multivariate normal integral. Biometrika, 57, 657-660.

X is MVN with mean zero and variance V. The probability of $\{X > a\}$. Modification of Das' method and generalization by expressing $V = C^2+BB'$. Sufficient condition for the representation. Bounds on the probability of $\{X > a\}$ by use of Slepian's result. Example.

1160. Wegner, P. (1963): Relations between multivariate statistics and mathematical programming. Appl. Statist., 12, 146-150.

Similarities and differences between multivariate analysis and mathematical programming. Need for interdisciplinary approach.

1161. Weiler, H. (1959): Means and standard deviations of a truncated normal bivariate distribution. Austral. J. Statist., 1, 73-81.

Truncated BVN with single truncation in both variables. First and second mements. Tables and construction and use of charts.

Weiler, H. (1964): See Williams, J.M. and Weiler, H., #1176.

1162. Weiner, J.M. and Dunn, O.J. (1966): Elimination of variates in linear discrimination problems. Biometrics, 22, 268-275.

Misclassification probabilities. Method of selecting variables for inclusion in a linear discriminant function: t-statistic, discriminant function coefficients, stepwise regression and random selection. Comparison by the probabilities of misclassification. Methods applied to data.

1163. Weingarten, H. and Di Donato, A.R. (1961): A table of generalized circular error. Math. Comput., 15, 169-173.

BVN and the probabilities of circular regions. Computing schemes. Tables of values of k satisfying $P\{X^2+Y^2 \leq k^2\sigma_x^2\} = \alpha$ *for the circular case. Application and statistical interpretation.*

Weitzman, R.A. (1964): See Ross, J. and Weitzman, R.A., #957.

1164. Welch, P.D. and Wimpress, R.S. (1961): Two multivariate statistical computer programs and their application to the vowel recognition problem. J. Acoust. Soc. Amer., 33, 426-434.

Classification into k classes using a Bayesian procedure and Anderson-Bahadur rule for unequal variance case. Stepwise application. Probabilities of misclassification. Application to vowel recognition problems.

Wette, R. (1972): See Choi, S.C. and Wette, R., #165.

1165. Wherry, R.J., Naylor, J.C., Wherry, R.J. Jr. and Fallis, R.F. (1965): Generating multiple samples of multivariate data with arbitrary population parameters. Psychometrika, 30, 303-313.

A method for computer generation of any number of scores and correlation matrices. Conversion to any predetermined covariance structure discussed. Example.

lliams, J.S. (1972): See Wu, S., Williams, J.S. and Mielke, P.W. Jr., #1180.

mpress, R.S. (1961): See Welch, P.D. and Wimpress, R.S., #1164.

lfe, J.H. (1970): Pattern clustering by multivariate mixture analysis. Multivariate Behav. Res., 5, 329-350.

Cluster analysis formulated as a problem of estimating the parameters of a mixture of multivariate distributions. ML estimation. Examples in which dispersion matrices are equal and unequal.

, S., Williams, J.S. and Mielke, P.W. Jr. (1972): Some designs and analyses for temporally independent experiments involving correlated bivariate responses. Biometrics, 28, 1043-1061.

Designs and analyses for experiments with two experimental units. Responses correlated within and independent among time periods. Maximization of Fisher's information function. Motivated by cloud seeding experiments.

le, C. (1972): See Jenden, D.J., Fairchild, M.D., Mickey, M.R., Silverman, R.W. and Yale, C., #513.

o, Y. (1965): An approximate degrees of freedom solution to the multivariate Behrens-Fisher problem. Biometrika, 52, 139-147.

Behrens-Fisher problem. Test of $\mu_1 = \mu_2$ when $\Sigma_1 \neq \Sigma_2$. Approximate degrees of freedom techniques. James' and Bennett's procedures. Monte Carlo methods. Generation of Wishart matrices. Representation of level of significance as a conditional expectation. Table comparing exact α and nominal $\alpha = .05$ for BVN case.

h, N. (1968): A multivariate normal test with two-sided alternative. Bull. Math. Statist., 13, 85-88.

Test on the mean vector when Σ is known. Two-sided alternatives with strict inequality on at least one component. Distribution of LR criterion. Tables of percentage points.

oneda, K. (1961): Some estimations of the parameters of multinormal populations from linearly truncated samples I. Yokohama J. Math., 9, 149-161.

Definition of linear truncation. Linear truncation in multivariate case. Reduction to bivariate case. Moments of linearly truncated bivariate normal. Estimation of parameters using large samples. Method of moments estimation.

Wherry, R.J. Jr. (1965): See Wherry, R.J., Naylor, J.C., Wherry, R.J. Jr. and Fallis, R.F., #1165.

Wiesen, J.M. (1959): See Owen, D.B. and Wiesen, J.M., #859.

1166. Wigner, E.P. (1967): Random matrices in physics. SIAM Rev., 9, 1-23.

Review paper emphasizing the role of characteristic roots in problems occurring in physics. Work of physicists in this context discussed to greater extent than that of statisticians. Interesting physical concepts introduced.

1167. Wijsman, R.A. (1957): Random orthogonal transformations and their use in some classical distribution problems in multivariate analysis. Ann. Math. Statist., 28, 415-423.

Distribution theory. Transformations using random orthogonal matrices. Application of technique. Derivation of distributions of T^2, generalized variance, Wishart and Bartlett's decomposition.

1168. Wijsman, R.A. (1959): Applications of a certain representation of the Wishart matrix. Ann. Math. Statist., 30, 597-601.

Representation of $W = CTT'C'$ when $CC' = \Sigma$. Recognition of the distribution of the elements of T. Application to (i) distribution of diagonal elements of A^{-1} and hence T^2, (ii) sample correlation and multiple correlation coefficients, (iii) characteristic roots and sphericity criterion for $p = 2$.

1169. Wilk, M.B. and Gnanadesikan, R. (1961): Graphical analysis of multiresponse experimental data using ordered distances. Proc. Nat. Acad. Sci. U.S.A., 47, 1209-1212.

Graphical procedures. Extension of Daniel's half-normal plotting techniques to multivariate data. Distances based on contrast vectors and compounding matrices. Order statistics in a gamma population.

1170. Wilk, M.B. and Gnanadesikan, R. (1964): Graphical methods for internal comparison in multiresponse experiments. Ann. Math. Statist., 35, 613-631.

The 2^n factorial. Graphical analyses vs. multivariate statistical analyses. Discussion of graphical procedures; uses and abuses. Choice of various arbitrary factors in the analysis. Examples. Graphical display of results. General discussion. Gamma and half-normal plotting.

Wilk, M.B. (1969): See Gnanadesikan, R. and Wilk, M.B., #381.

1171. Wilks, S.S. (1960): Multidimensional statistical scatter. Contrib. Prob. Statist., (Hotelling Volume, Olkin et al eds.), 486-503.

An expository article giving the basic notions of scatter in the one, two and k dimensions. Minimization of scatter and expected value of the sample scatter. Internal scatter. Multinormal case. Relationship of the k-dimensional scatter with Hotelling's T^2, principal components. Discriminant analysis (two and several sample cases), canonical correlations. Scatter ratios and their generalizations. An optimality of principal components in respect of scatter.

1172. Wilks, S.S. (1963): Multivariate statistical outliers. Sankhya, A, 25, 407-426.

Use of scatter ratios to test for outliers: (i) one outlier case: $R_1,\ldots,R_n$ where R_j is the ratio of the internal scatter of the jth observation to that of the total sample. Ordered set of scatter ratios. Test for one outlier based on $r_1 = \min(R_1,\ldots,R_n)$ and rejection of an observation for small values. Probability bounds. Critical values of r_1 based on the bounds. Tables. (ii) Two outlier case: scatter ratio in this case R_{ij} = ratio of the internal scatter by deleting two observations to the total internal scatter in a sample. Order set of R_{ij} and $r_2 = \min\{R_{ij}: i < j\}$. Deletion of observations based on r_2. Null distribution and moments. Approximate critical values and tables. (iii) Case of more than two outliers considered briefly.

1173. Williams, E.J. (1961): Tests for discriminant functions. J. Austral. Math Soc., 2, 243-252.

Discriminant and classificatory problems. Adequacy of a single discriminant function. Linear relations and discriminant functions. Tests for direction of an explanatory variable. Relationship with regression analysis. Case of several explanatory variabes. Group differences. Distributional problems and large sample results. MV analysis of covariance.

1174. Williams, E.J. (1967): The analysis of
variates. J. Roy. Statist. Soc.

*An expository paper detailing the a
data, with a discussion of computer
mination, canonical correlation, te
minators, residual (adjusted) discr
Tests for coplanarity, direction an
Tests in terms of canonical variate
to discrete data. Tests for additi
Large sample theory. A discussion*

1175. Williams, E.J. (1970): Comparing means
trika, 57, 459-461.

*Generalization of the test for axial
sideration of the hypothesis H_0: ε'
any vector. Construction of Anders
function of ε and T the residual ma
unknown, Σ known or $\Sigma = \sigma^2 k$ where k
Application to the model $\mu = \delta+\lambda\varepsilon$.*

1176. Williams, J.M. and Weiler, H. (1964): Fu
truncated normal bivariate distri
6, 117-129.

*Rectangular truncation in BVN, of th
$b \leq Y < \infty$. Charts for E(X), E(Y) an
population retained after truncatio
for various a and b.*

1177. Williams, J.S. (1967): The variance of w
J. Amer. Statist. Assoc., 62, 129

*Weighted lease squares estimation wh
and error has MVN structure. Respon
neralized variance. Measure of loss
to estimation of Σ. Wilks' MANOVA c*

1178. Williams, J.S. (1970): The choice and us
of two sets of variates. Biometr

*NSC for X'Y and $X_c'Y$ to be independen
Linear transformation of principal c
to determine if a set of vectors X'Y
formation on the principal component
characterization. Tests of independ
to growth curves, to separate the ge
the post-prenatal environmental effe*

Wi

Wi

1179. Wo

1180. Wu

Ya

1181. Ya

1182. Y

1183. Y

1184. Young, D.H. (1967): Recurrence relations between the PDF's of order statistics of dependent variables and some applications. Biometrika, 54, 283-292.

General formulae for moments and p.d.f.'s: recurrence relationships. Application to equicorrelated normal variables and F-ratios. Approximation to upper percentage points of distribution of: (i) extreme order statistics, (ii) second largest order statistic. Tables.

Young, D.L. (1971): See Pillai, K.C.S. and Young, D.L., #902.

Young, D.L. (1971): See Pillai, K.C.S. and Young, D.L., #903.

1185. Young, J.C. (1971): Some inference problems associated with the complex multivariate normal distribution. Tech. Rep. #102, Dept. Statist., Southern Methodist U.

Complex analogues of problems associated with real MVN distribution. Bartlett's decomposition of a complex Wishart matrix. Coherence (correlation) between two complex random variables. Multiple coherence. Test for equality of means. Test for dimensionality of mean space. Goodness-of-fit for a single hypothetical discriminant function.

Young, S.S.Y. (1962): See Tallis, G.M. and Young, S.S.Y., #1117.

Zaatar, M.K. (1972): See Srivastava, J.N. and Zaatar, M.K., #1073.

Zakharov, V.K. (1960): See Sarmanov, O.V. and Zakharov, V.K., #993.

1186. Zaslavskii, A.E. (1967): On statistical tests for a simple hypothesis involving a multidimensional normal distribution. Theory Prob. Appl., 12, 514-519.

Tests of $\mu = \mu_1$ vs. $\mu = \mu_2$ in MVN case when Σ is replaced by Σ_0, a diagonal matrix. Effect of replacing Σ by Σ_0 on probability of Type II error when (i) parameters are known and (ii) parameters are estimated. Bounds on probability.

1187. Zelen, M. and Severo, N.C. (1960): Graphs for bivariate normal probabilities. Ann. Math. Statist., 31, 619-624.

Charts for computing BVN probability integral. Maximum error of 10^{-2}. Motivated by work of Owen and Wiesen. Applications.

Zellner, A. (1961): See Hooper, J.W. and Zellner, A., #473.

Zellner, A. (1964): See Tiao, G.C. and Zellner, A., #1125.

1188. Zellner, A. and Chetty, V.K. (1965): Prediction and decision problems in regression models from the Bayesian point of view. J. Amer. Statist. Assoc., 60, 608-616. [Correction: 63 (1968), 1551].

Bayesian prediction in multivariate regression model. Posterior distribution of a future vector of observations. Geisser-Cornfield prior. Determination of X for future vector so as to minimize expected quadratic loss function.

1189. Zhurbenko, I.G. (1968): Moments of random determinants. Theory Prob. Appl., 13, 682-686.

Moments of determinant Δ_n: $E[\Delta_n^2]$, $E[\Delta_n^{2k}]$, $E[\Delta_n^{2k+1}]$. *Recurrence relations among the moments with respect to dimension on determinant. Generating functions. Asymptotic expansions.*

LIST OF ARTICLES NOT ABSTRACTED

Chang, T.C. (1969): See Pillai, K.C.S. and Chang, T.C. (1969).

Cheng, M.C. (1970): Some properties of the incomplete moments of the multivariate distribution. Nanta Math., 3 (II), 84-87.

Clunies-Ross, C.W. (1960): See Riffenburgh, R.H. and Clunies-Ross, C.W. (1960).

Eidemiller, R.L. (1964): See Lipow, M. and Eidemiller, R.L. (1964).

Harris, D.L. (1963): See Hartley, H.O. and Harris, D.L. (1963).

Hartley, H.O. and Harris, D.L. (1963): Monte Carlo computations in normal correlation problems. J. Assoc. Comput. Mach., 10, 302-306.

Lipow, M. and Eidemiller, R.L. (1964): Application of bivariate normal distribution to a stress vs. strength problem in reliability analysis. Technometrics, 6, 325-328.

Mandel, J. (1969): The partitioning of interaction in analysis of variance. J. Res. Nat. Bur. Standards, Sect. B, 73, 309-328.

Mathai, A.M. (1970): The exact distribution of a criterion for testing the hypothesis that several multivariate populations are identical. J. Indian Statist. Assoc., 8, 1-17.

Miller, K.S. and Sackrowitz, H. (1965): Distributions associated with the quadrivariate normal. Indust. Math., 15, 1-14.

Moustafa, M.D. (1957): A note on partial correlation. Egyptian Statist. J., 1, 1-5.

Moustafa, M.D. (1958): Testing of hypotheses on a multivariate population; some of the variates being continuous and the rest categorical. Egyptian Statist. J., 2, 73-96.

Nagao, H. (1970): Asymptotic expansions of some test criteria for homogeneity of variances and covariance matrices from normal populations. J. Sci. Hiroshima Univ. Ser. A-1, 34, 153-247.

Pillai, K.C.S. and Chang, T.C. (1969): An approximation to the c.d.f.

of the largest root of a covariance matrix. Ann. Inst. Statist. Math., 6, (Supplement), 115-124.

Rahman, N.A. (1963): On the sampling distribution of the Studentized Penrose measure of distance. Ann. Human Genet., 26, 97-106.

Riffenburgh, R.H. and Clunies-Ross, C.W. (1960): Linear discriminant analysis. Pacific Sci., 14, 251-256.

Roux, J.J.J. (1971): On generalized multivariate distributions. South African Statist. J., 5,

Sackrowitz, H. (1965): See Miller, K.S. and Sackrowitz, H. (1965).

Siotani, M. (1958): Note on the utilization of the generalized student ratio in the analysis of variance or dispersion. Ann. Inst. Statist. Math., 9, 157-171.

Tallis, G.M. (1967): Approximate maximum likelihood estimates from grouped data. Technometrics, 9, 599-606.

Tan, W.Y. (1968-69): Some results of multivariate regression analysis. Nanta Math., 3, 54-71.

Tan, W.Y. (1968-69): On the complex analogue of Bayesian estimation of a multivariate regression model. Nanta Math., 3, 72-99.

Tukey, J.W. and Wilk, M.B. (1966): Data analysis and statistics, an expository overview. Fall. Joint Comp. Conf. Proc., 29, 695-710.

Tumura, Y. (1965): The distribution of latent roots and vectors. T.R.U. Math., 1, 1-16.

Wilk, M.B. (1966): See Tukey, J.W. and Wilk, M.B. (1966).

Yoshimura, I. (1964): Unified system of cumulant recurrence relations. Rep. Statist. Appl. Res. JUSE, 11, 1-8.

Yoshimura, I. (1964): A complementary note on the multivariate moment recurrence relation. Rep. Statist. Appl. Res. JUSE, 11, 9-12.

LIST OF JOURNALS

Acad. Roy. Belg. Bull. Cl. Sci.: *Academie Royale de Belgique: Bulletine de la Classe des Sciences, 5eme Serie.*

Amer. Math. Monthly: *The American Mathematical Monthly.*

Ann. Human Genet.: *Annals of Human Genetics.*

Ann. Inst. Statist. Math.: *Annals of the Institute of Statistical Mathematics.*

Ann. Math. Ser. 2: *Annals of Mathematics, Series 2.*

Ann. Math. Statist.: *The Annals of Mathematical Statistics.*

Appl. Statist.: *Applied Statistics*

Arch. Math.: *Archiv der Mathematik.*

Austral. J. Statist.: *The Australian Journal of Statistics.*

Bell System Tech. J.: *The Bell System Technical Journal.*

Biometrics: *Biometrics.*

Biometrika: *Biometrika.*

British J. Math. Statist. Psychol.: *British Journal of Mathematical and Statistical Psychology.*

Bull. Calcutta Math. Soc.: *Bulletin of the Calcutta Mathematical Society.*

Bull. Inst. Internat. Statist.: *Bulletin de l'Institut International de Statistique.*

Bull. Math. (Romanian): *Bulletin Mathematique.*

Bull. Math. Statist.: *Bulletin of Mathematical Statistics.*

Calcutta Math. Soc. Golden Jubilee Commem. Volume: *Calcutta Mathematical Society Golden Jubilee Commemorative Volume.*

Calcutta Statist. Assoc. Bull.: *Calcutta Statistical Association Bulletin.*

Canad. Math. Bull.: *Canadian Mathematical Bulletin.*

Contrib. Order Statistics (Sarhan and Greenberg, eds.): *Contributions to Order Statistics. Editors: A. E. Sarhan and B. G. Greenberg.*

Contrib. Prob. Statist. (Hotelling Volume, Olkin et al, eds.): *Contributions to Probability and Statistics. Essays in honour of H. Hotelling. Editors: I. Olkin, S. G. Ghurye, W. Hoeffding, W. G. Madow, and H. B. Mann.*

Contrib. Statist. (Mahalanobis Volume, Rao, ed.): *Contributions to Statistics, presented to Professor P. C. Mahalanobis on the occasion of his 70th birthday. Editor: C. R. Rao.*

CSIRO Div. Math. Statist. Tech. Paper: *Commonwealth Scientific and Industrial Research Organization, Division of Mathematical Statistics, Technical Paper.*

Duke Math. J.: *Duke Mathematical Journal.*

Econometrica: *Econometrica.*

Egyptian Statist. J.: *Egyptian Statistical Journal.*

Essays Prob. Statist. (Roy Volume, Bose et al, eds.): *Essays on Probability and Statistics. In memory of S. N. Roy. Editors: R. C. Bose, I. M. Chakravarti, P. C. Mahalanobis, C. R. Rao and K. J. C. Smith.*

Fall Joint Comp. Conf. Proc.: *Fall Joint Computer Conferences Proceedings.*

Growth: *Growth.*

IEEE Trans. Information Theory: *IEEE Transactions on Information Theory.*

Indust. Math.: *The Journal of the Industrial Mathematics Society.*

Indust. Qual. Control: *Industrial Quality Control.*

IRE Trans. Electronic Comp.: *IRE Transactions on Electronic Computers.*

IRE Trans. Information Theory: *IRE Transactions on Information Theory.*

J. Acoust. Soc. Amer.: *The Journal of the Acoustical Society of America.*

J. Amer. Statist. Assoc.: *Journal of the American Statistical Association.*

J. Appl. Prob.: *Journal of Applied Probability.*

J. Assoc. Comput. Mach.: *Journal of the Association for Computing Machinery.*

J. Austral. Math. Soc.: *The Journal of the Australian Mathematical Society.*

J. Chronic Diseases: *Journal of Chronic Diseases.*

J. Gerontology: *Journal of Gerontology.*

J. Indian Soc. Agric. Statist.: *Journal of the Indian Society of Agricultural Statistics.*

J. Indian Statist. Assoc.: *Journal of Indian Statistical Association.*

J. Multivariate Anal.: *Journal of Multivariate Analysis.*

J. Opt. Soc. Amer.: *Journal of the Optical Society of America.*

J. Res. Nat. Bur. Standards Sect. B: *Journal of Research of the National Bureau of Standards, Section B.*

J. Roy. Statist. Soc., A: *Journal of the Royal Statistical Society, Series A.*

J. Roy. Statist. Soc., B: *Journal of the Royal Statistical Society, Series B.*

J. Sci. Hiroshima Univ. Ser. A-1: *Journal of Science of the Hiroshima University, Series A.*

Kumamoto J. Sci. Ser. A: *Kumamoto Journal of Science, Series A.*

Math. Comput.: *Mathematics of Computation.*

Math. Nachr.: *Mathematische Nachrichten.*

Mem. Fac. Sci. Kyushu Univ. Ser. A: *Memoirs of the Faculty of Science, Kyushu University, Series A.*

Meteorol. Monog.: *Meteorological Monographs.*

Metrika: *Metrika.*

Metron: *Metron.*

Multivariate Anal. I (Proc. 1st Internat. Symp., Krishnaiah, ed.): *Proceedings of the First International Symposium on Multivariate Analysis. Editor: P. R. Krishnaiah.*

Multivariate Anal. II (Proc. 2nd Internat. Symp., Krishnaiah, ed.): *Proceedings of the Second International Symposium on Multivariate Analysis. Editor: P. R. Krishnaiah.*

Multivariate Behav. Res.: *Multivariate Behavioral Research.*

Nanta Math.: *Nanta Mathematica.*

Nuclear Phys.: *Nuclear Physics.*

Operations Res.: *Operations Research.*

Osaka J. Math.: *Osaka Journal of Mathematics.*

Osaka Math. J.: *Osaka Mathematical Journal.*

Pacific Sci.: *Pacific Science.*

Proc. Cambridge Philos. Soc.: *Proceedings of the Cambridge Philosophical Society.*

Proc. Fifth Berkeley Symp. Math. Statist. Prob.: *Proceedings of the Fifth Berkeley Symposium on Mathematical Statistics and Probability.*

Proc. Fourth Berkeley Symp. Math. Statist. Prob.: *Proceedings of the Fourth Berkeley Symposium on Mathematical Statistics and Probability.*

Proc. IBM Sci. Comput. Symp. Statist.: *Proceedings of the IBM Scientific Computing Symposium.*

Proc. Nat. Acad. Sci. U.S.A.: *Proceedings of the National Academy of Sciences of USA.*

Proc. Roy. Soc. Edinburgh Sect. A: *Proceedings of the Royal Society of Edinburgh, Section A.*

Proc. Roy. Soc. London Ser. A: *Proceedings of the Royal Society of London, Series A.*

Proc. Sixth Berkeley Symp. Math. Statist. Prob.: *Proceedings of the Sixth Berkeley Symposium on Mathematical Statistics and Probability.*

Psychol. Bull.: *Psychological Bulletin.*

Psychomet. Monog.: *Psychometric Monographs.*

Psychometrika: *Psychometrika.*

Quart. Appl. Math.: *Quarterly of Applied Mathematics.*

Rep. Statist. Appl. Res. JUSE: *Reports of Statistical Applications Research. Union of Japanese Scientists and Engineers.*

Res. Papers Statist. Festschr. Neyman (David, ed.): *Research Papers in Statistics. Festschrift for J. Neyman. Editor: F. N. David.*

Rev. Inst. Internat. Statist.: *Review of the International Statistical Institute.*

S. African J. Agri. Sci.: *South African Journal of Agricultural Science.*

South African Statist. J.: *South African Statistical Journal.*

Sankhya: *Sankhya.*

Sankhya, A: *Sankhya, Series A.*

Sankhya, B: *Sankhya, Series B.*

SIAM J. Appl. Math.: *SIAM Journal on Applied Mathematics.*

SIAM Rev.: *SIAM Review.*

Soviet Math. Dokl.: *Soviet Mathematics - Doklady.*

Statistica: *Statistica.*

Statistica Neerlandica: *Statistica Neerlandica.*

Statistician: *The Statistician.*

Stud. Item Anal. Predict. (Solomon, ed.): *Studies in Item Analysis and Prediction. Editor: Herbert Solomon.*

T.R.U. Math.: *Journal of Mathematics, Tokyo Rica University.*

Technometrics: *Technometrics.*

Theor. Math. Biol. (Waterman and Morowtiz, eds.): *Theoretical and Mathematical Biology. Editor: Talbot H. Waterman and Harold J. Morowitz.*

Theory Prob. Appl.: *Theory of Probability and Its Applications.*

Trabajos Estadist.: *Trabajos de Estadistica.*

Ukranian Math. J.: *Ukranian Mathematical Journal.*

Yokohama J. Math.: *Yokohama Journal of Mathematics.*

SUBJECT INDEX

Numbers refer to the abstracts